SpringerBriefs in Astronomy

Series Editors

Martin Ratcliffe
Wolfgang Hillebrandt
Michael Inglis

For further volumes:
http://www.springer.com/series/10090

Laura Brenneman

Measuring the Angular Momentum of Supermassive Black Holes

 Springer

Laura Brenneman
Smithsonian Astrophysical Observatory
Cambridge, MA, USA

ISSN 2191-9100   ISSN 2191-9119 (electronic)
ISBN 978-1-4614-7770-9   ISBN 978-1-4614-7771-6 (eBook)
DOI 10.1007/978-1-4614-7771-6
Springer New York Heidelberg Dordrecht London

Library of Congress Control Number: 2013941131

© Laura Brenneman 2013
This work is subject to copyright. All rights are reserved by the Publisher, whether the whole or part of the material is concerned, specifically the rights of translation, reprinting, reuse of illustrations, recitation, broadcasting, reproduction on microfilms or in any other physical way, and transmission or information storage and retrieval, electronic adaptation, computer software, or by similar or dissimilar methodology now known or hereafter developed. Exempted from this legal reservation are brief excerpts in connection with reviews or scholarly analysis or material supplied specifically for the purpose of being entered and executed on a computer system, for exclusive use by the purchaser of the work. Duplication of this publication or parts thereof is permitted only under the provisions of the Copyright Law of the Publisher's location, in its current version, and permission for use must always be obtained from Springer. Permissions for use may be obtained through RightsLink at the Copyright Clearance Center. Violations are liable to prosecution under the respective Copyright Law.
The use of general descriptive names, registered names, trademarks, service marks, etc. in this publication does not imply, even in the absence of a specific statement, that such names are exempt from the relevant protective laws and regulations and therefore free for general use.
While the advice and information in this book are believed to be true and accurate at the date of publication, neither the authors nor the editors nor the publisher can accept any legal responsibility for any errors or omissions that may be made. The publisher makes no warranty, express or implied, with respect to the material contained herein.

Printed on acid-free paper

Springer is part of Springer Science+Business Media (www.springer.com)

Contents

Chapter 1
Introduction

Black holes represent the ultimate frontier in astrophysics: the one-way passage to the unknown and the unknowable. These exotic objects are defined by a characteristic radius known as the event horizon: the radius from the central collapsed remnant (or singularity) at which the escape velocity of the black hole equals the speed of light. Black holes therefore emit no light themselves, and we can only observe them indirectly by analyzing the electromagnetic (e/m) radiation released from the gas they accrete. This accretion typically takes the form of a geometrically-thin, optically-thick disk (Shakura and Sunyaev 1973) for black holes which are actively accreting gas ($L_X/L_{Edd} \geq 0.001$, Miller 2007). The finite value of the speed of light renders all material and spacetime within the event horizon causally separated from the Universe in which we live; at present, there is no known way to access information from beyond the event horizon.[1] Due to this limitation, all of our knowledge of black hole systems comes from Einstein's Special and General theories of Relativity, and from e/m observations of accretion disks around known or suspected black hole systems which are, almost invariably, bright and/or nearby.

In spite of their enigmatic nature, black holes are arguably the simplest objects in the Universe, possessing only three fundamental properties by which they can be completely defined: (1) mass, (2) spin, and (3) electric charge. In practice, the electric charge of a black hole in any environment other than a pure vacuum is assumed to be negligible, as the black hole would rapidly accrete oppositely charged particles and neutralize itself. Mass and spin—or angular momentum—are thus the only two meaningful properties that describe an astrophysical black hole (Kerr 1963). The mass of a black hole determines the degree to which the spacetime in

---

[1] Quantum mechanical "tunneling" theoretically enables black holes to emit thermal radiation at a very slow rate. This is known as "Hawking Radiation," (Hawking 1974) and can eventually evaporate a black hole. However, a black hole with solar mass would take $\sim 10^{66}$ years to evaporate via this process. A supermassive black hole would take considerably longer.

L. Brenneman, *Measuring the Angular Momentum of Supermassive Black Holes*,
SpringerBriefs in Astronomy, DOI 10.1007/978-1-4614-7771-6_1,
© Laura Brenneman 2013

which it resides is warped (as in the classic "bowling ball on a trampoline" analogy), whereas spin determines the degree to which that spacetime is twisted (much like beaters in thick batter).

The masses of stellar-mass black holes within our own galaxy are typically determined by examining the orbital and radiative properties of their companion stars. Measuring the masses of SMBHs can be more difficult, however. Several methods have been developed to estimate the masses of SMBHs: e.g., reverberation mapping (Blandford and McKee 1982), stellar velocity dispersion (Ferrarese and Merritt 2000), tracing of stellar orbits (Genzel et al. 2000; Ghez et al. 2000), maser observations (Watson and Wallin 1994), and gravitational lensing (Silvestro 1974). Most of these methods rely on measuring radiation emitted relatively far from the black hole. Observationally, black holes range in size from $\sim$3 to $10^{10}\,M_\odot$, with most stellar-mass black holes clustered in the 5–20 $M_\odot$ range and most SMBHs with masses of 10^6–$10^8\,M_\odot$. Over the last decade there has been some evidence to support the existence of intermediate mass black holes with masses of order 10^2–$10^4\,M_\odot$ (e.g., Miller and Colbert 2004).

Though black hole mass is by no means trivial to measure, spin is the more challenging property to ascertain. In contrast to constraining mass, measuring spin requires probing the nature of the spacetime within a few gravitational radii of the event horizon (where the gravitational radius is defined as $r_g \equiv GM/c^2$; G is Newton's constant, M is the mass of the black hole and c is the speed of light). The angular momentum of a black hole only manifests through Lense-Thirring precession, also known as frame-dragging, which occurs only in the innermost part of the accretion disk where relativistic effects cause the spacetime in this region to become twisted in the same direction that the black hole is rotating. To observe this effect, observations of the innermost disk must be made in X-rays, given the energetic processes at work in the cores of actively-accreting black holes. Current X-ray telescopes lack the spatial resolution necessary to resolve the innermost regions of the accretion disk, even in bright, nearby AGN. As such, X-ray spectra taken from the core of the AGN are the tool of choice for examining the spacetime of the inner disk.

Spin (in dimensionless form) is defined as $a \equiv cJ/GM^2$, where cosmic censorship within the framework of General Relativity dictates that $-1 \leq a \leq +1$ (negative spin values represent retrograde configurations in which the black hole spins in the opposite direction to the disk, positive values denote prograde spin configurations, and $a = 0$ implies a non-spinning black hole), and J represents the black hole angular momentum (Bardeen et al. 1972; Thorne 1974). If spin is known to within $\Delta a \leq 10\%$, then meaningful correlations can be drawn between spin and other environmental variables, e.g., the history of the accretion flow and the presence and power of relativistic jets in the system.

Supermassive black holes are particularly interesting to examine, given that their masses and spins have likely evolved considerably in the billions of years since their formation. SMBHs grow by either merging with other black holes or accreting gas, most often by a combination of the two processes (e.g., Volonteri et al. 2005). Additionally, as a SMBH grows, it can go through periods where it produces powerful outflows of kinetic and radiative energy through the production of winds and jets (Fabian 2012), seeding the surrounding environment with matter and energy. Such heating of the ambient gas in and around the host galaxy may ultimately play a significant role in regulating its rate of star formation. This type of "feedback" process has been cited as a potential explanation for the famed $M - \sigma$ relation linking the mass of the SMBH to the velocities of the stars in the central bulge of its host galaxy, as well as to the mass of the bulge itself (e.g., Ferrarese and Merritt 2000; Gültekin et al. 2009). Given that jets are thought to be launched by the magnetic extraction of rotational energy from the ergosphere of the black hole (Blandford and Znajek 1977) when the black hole spin gets sufficiently large ($a \geq 0.93$; Agol and Krolik 2000), spin may play a significant role in regulating galaxy growth on scales far beyond the gravitational sphere of influence of the black hole.

Put simply, measurements of the spins of SMBHs in AGN can contribute to our understanding of these complex and energetic environments in three principal ways:

- They offer a rare probe of the nature of the spacetime proximal to the event horizon of the black hole, well within the strong-field gravity regime (Fabian et al. 1989; Laor 1991);
- They can shed light on the relation of a black hole's angular momentum to its outflow power in the form of jets (e.g., Narayan and McClintock 2012; Steiner et al. 2012 for stellar-mass black holes);
- They can also inform us about the relative role of gas accretion vs. mergers in recent epochs of the life of the host galaxy and its AGN (Berti and Volonteri 2008).

For these reasons, developing a theoretical and observational framework in which to measure black hole spin accurately and precisely is of critical importance to our understanding of how galaxies form and evolve over cosmic time.

Advances in theoretical modeling as well as observational sensitivity in the *Chandra/XMM-Newton/Suzaku* era are finally producing robust constraints on the spins of a handful of SMBHs. Computationally, new algorithms developed within the past decade (Dovčiak et al. 2004; Beckwith and Done 2005; Brenneman and Reynolds 2006; Dauser et al. 2010, 2013) have made it possible to perform fully relativistic ray-tracing of photon paths emanating from the accretion disk close to the black hole, keeping the black hole spin as a variable parameter in the model. When such models are fit to high signal-to-noise (S/N) X-ray spectra from the innermost accretion disk, they yield vital physical information about the black hole/disk system, including constraints on how fast—and in what direction—the black hole is rotating.

In this work, I discuss our current knowledge of the distribution of SMBH spins in the local universe and future directions of black hole spin research. I begin in Chap. 2 with an examination of the spectral modeling techniques used to measure black hole spin, focusing on those most effective in constraining spin in AGN. I then discuss the models involved, reviewing the caveats that must be considered in the process in Chap. 3. In Chap. 4 I demonstrate the application of these techniques to deep observations of the nearby, type 1 AGN MCG–6-30-15, NGC 3783 and Fairall 9. I examining our current knowledge of the spin distribution of local SMBHs in Chap. 5, along with its implications. Future directions for this field of research are presented in Chap. 6.

Chapter 2
Measuring Black Hole Spin

In principle, there are at least five ways that spin can be measured for a single (i.e., non-merging) SMBH. All five are predicated on the assumption that General Relativity provides the correct description of the spacetime near the black hole, and that there is an easily-characterized, monotonic relation between the radius of the innermost stable circular orbit (ISCO) of the accretion disk and the black hole spin (see Fig. 2.1). The disk is also assumed to remain geometrically thin, optically thick and radiatively efficient down to the ISCO boundary, and to truncate relatively rapidly therein (see Sect. 5.1).

The five methods for measuring spin are listed below.

- **Thermal Continuum Fitting** (e.g., Remillard and McClintock 2006) treats the inner accretion disk as a modified blackbody, and the radius of the ISCO is computed via the Stefan-Boltzmann law, by measuring the peak temperature and flux of this blackbody ($Fd^2 \propto R^2 T^4 \cos(i)$, where F is the disk blackbody flux, d is the distance to the source, R is the radius of the ISCO, T is the peak blackbody temperature of the disk, and i is the inclination angle of the disk to our line of sight). The physics behind this method is straightforward, much like the method one would use to measure the photospheric radius of a star. When dealing with an accretion disk, however, caveats include the degree to which the disk emission is Comptonized by the highly ionized disk atmosphere ("spectral hardening," as per Davis et al. 2006), which can be difficult to quantify precisely. The disk luminosity must also fall within a range roughly 1–30% of the Eddington value in order to ensure that the blackbody emission dominates over the Comptonized, power-law component. Because this method also relies on precise, accurate, independent measurements of the distance to and mass of the black hole, as well as its disk inclination angle, the thermal continuum fitting method can only viably be used to measure spin in stellar-mass black holes (for which there are 14 published spin constraints at the time of this writing). Moreover, the temperature of the accretion disk goes as $T \propto M^{-1/4}$ (Frank, King and Raine 2002), so the blackbody peak for AGN disks is in the UV,

L. Brenneman, *Measuring the Angular Momentum of Supermassive Black Holes*,
SpringerBriefs in Astronomy, DOI 10.1007/978-1-4614-7771-6_2,
© Laura Brenneman 2013

whereas stellar-mass black hole disks peak in soft X-rays. The prevalence of absorption in the UV band can present serious complications for accurately measuring the thermal disk emission in AGN.

- **Inner Disk Reflection Modeling** (e.g., Brenneman and Reynolds 2006; hereafter BR06) assumes that the high-energy X-ray emission (≥ 2 keV) is dominated by thermal, UV disk emission which has been Comptonized by hot electrons in a centrally-concentrated "corona." This structure may represent the disk atmosphere, the base of a jet or some alternative geometry (e.g., Markoff et al. 2005). Some of the Comptonized photons will scatter out of the system and form the power-law continuum characteristic of typical AGN in X-rays. A certain percentage of the photons, however, will be scattered back down ("reflected") onto the surface of the disk again. Provided that the disk is not completely ionized, the irradiation from the continuum power-law photons will excite a series of fluorescent emission lines of various atomic species at energies ≤ 7 keV, along with a "Compton hump" shaped by the Fe K absorption edge at ~ 7.1 keV and by downscattering at ~ 20–30 keV (see Fig. 2.2). The most prominent of the fluorescent lines produced is Fe Kα at a rest-frame energy of 6.4 keV, due largely to its high abundance and fluorescent yield. As such, the Fe Kα line is the most important diagnostic feature of the inner disk reflection spectrum; its shape is altered from the typical near-delta function profile expected in a laboratory, becoming highly broadened and skewed due to the combination of Doppler and relativistic effects close to the black hole (See Fig. 2.3). The energy at which the "red" wing (i.e., low-energy tail) of this line truncates is directly linked to the location of the ISCO, and therefore the spin of the black hole (see Reynolds and Nowak 2003 and Miller 2007 for comprehensive reviews of the reflection modeling technique). This method does not require *a priori* knowledge of the mass, distance or inclination of the black hole system in question, and can therefore be applied to any black hole system. However, the principal caveat for the reflection method is that the effects of disk ionization (especially in stellar-mass black holes with hotter disks, e.g., Davis et al. 2006) and/or complex absorption from gas along the line of sight to the system (particularly within AGN systems; e.g., Halpern 1984; Reynolds 1997) can make the determination of the low-energy bound of the red wing challenging.

- **High Frequency Quasi-Periodic Oscillations** (e.g., Strohmayer 2001), in which the X-ray power density spectrum of the emission from the inner accretion disk is characterized by 1–2 pulses at certain harmonic frequencies (e.g., in a 3:2 ratio), indicative of some type of resonance or regular oscillation within the accretion flow in this region. Such phenomena have been reported commonly in actively accreting stellar-mass black holes (e.g., Pétri 2008), but only once in an AGN (Gierlinski et al. 2008). The physical mechanism for producing these HFQPOs is not yet known, but if the frequencies at which they are observed are related in a fundamental way to the frequency of the ISCO (the Fourier transform of its period of rotation around the black hole), then the radius of the ISCO could potentially be derived from HFQPO observations, and the black hole spin thereby inferred.

- **X-ray Polarimetry** (e.g., Tomsick et al. 2009), in which the reflected emission from the inner accretion disk is expected to be polarized if the disk is indeed a geometrically-thin, optically thick slab of gas, as expected (i.e., Shakura and Sunyaev 1973). The degree and angle of the observed polarization vs. energy function would then have a characteristic shape depending on the spin of the black hole, due to the influence of the black hole spin on the shape of the spacetime immediately surrounding the black hole, and the radius of the ISCO (Schnittman and Krolik 2009). Measurements of the degree and angle of polarization vs. energy would require a sensitive X-ray polarimeter to be flown in space, however, and there are currently no active or planned missions to incorporate such an instrument.
- **Imaging the Event Horizon Shadow** (e.g., Broderick et al. 2011), in which sub-mm Very Long Baseline Interferometry (VLBI) is used to obtain micro-arcsecond spatial resolution to produce the first-ever images of the innermost accretion disk surrounding a black hole. By comparing these images with detailed models of the appearance of the innermost disk, which incorporate the necessary relativity and light bending as a function of physical parameters such as the radius of the ISCO, the black hole spin can be constrained (Doeleman et al. 2008). Both the modeling and VLBI measurement techniques are a work in progress at present, however, and this method is currently only able to achieve the spatial resolution necessary to image black holes with event horizons of comparatively large angular size (tens of micro-arcseconds), limiting the sample to Sgr A* and M87 at present.

There are limitations in applying the last three methods listed above, and the continuum fitting method has only been applied successfully to stellar-mass black holes. We are therefore currently restricted to using only the reflection modeling method for constraining the spins of SMBHs in AGN.

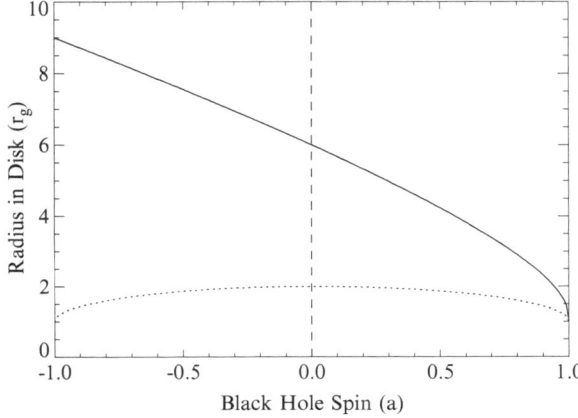

Fig. 2.1 Radius of the ISCO (*solid line*) and event horizon (*dotted line*) as a function of black hole spin. Spin values to the *left* of the *dashed line* indicate a black hole spinning in the opposite direction relative to the accretion disk (retrograde), while spins to the *right* of the *dashed line* indicate a black hole spinning in the same direction as the disk (prograde)

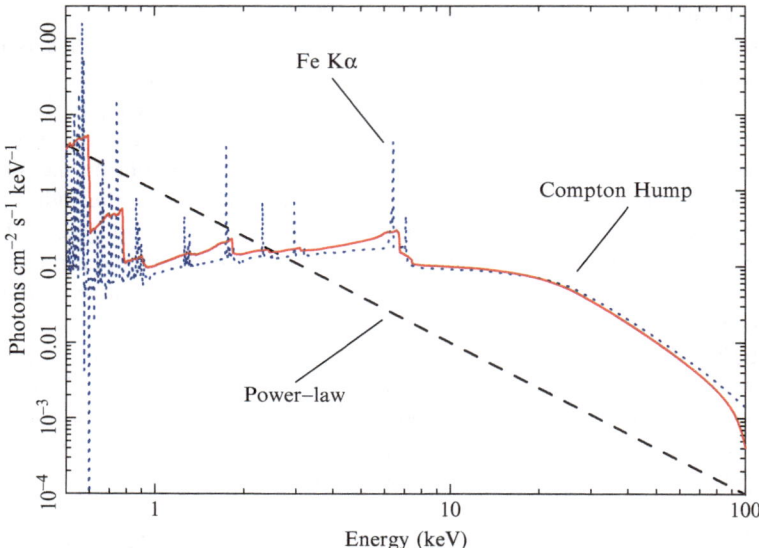

Fig. 2.2 The reflection spectrum for a neutral accretion disk with solar iron abundance, irradiated by a power-law continuum of $\Gamma = 2$. The power-law component is shown as a *dashed black line* while the static disk reflection spectrum is the *dotted blue line*. When the disk is convolved with Doppler and relativistic effects from the spacetime around even a non-spinning black hole extending down to the ISCO, we see the blurred spectrum in *solid red*. Note that the fluorescence lines are affected in shape more than the continuum, and that the Fe Kα line and Compton hump are especially prominent relative to the power-law. The static reflection spectrum is modeled by the `xillver` code of Garcia and Kallman (2010). The blurred reflection spectrum has been convolved with the `relconv` model of Dauser et al. (2010)

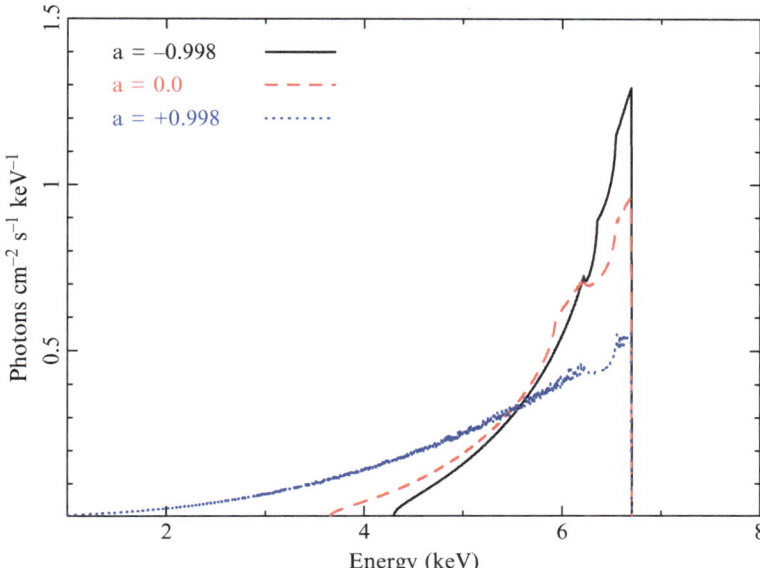

Fig. 2.3 Change in the shape of the Fe Kα line as a function of black hole spin. The *black solid line* represents $a = -0.998$, the *red dashed line* shows $a = 0$ and the *blue dotted line* shows $a = +0.998$. Note the enhancement in the breadth of the *red* wing of the line as the black hole spin increases. The `relline` code of Dauser et al. (2010) was used to plot the three lines

Chapter 3
Applying the Reflection Model

Though the reflection modeling method could, in principle, be used to measure the angular momentum of an actively accreting black hole of any mass, we restrict our focus here to determining spin in SMBHs.

3.1 Requirements

An AGN must satisfy a few important requirements in order to be a viable candidate for obtaining spin constraints. Firstly, it must be bright enough to achieve the necessary S/N in X-rays to accurately separate the reflection spectrum from (a) the continuum, and (b) any intrinsic absorption within the host galaxy and its nucleus. Typically one must obtain $\geq$200,000 photons over the 2–10 keV energy range (Guainazzi et al. 2006), though in practice the required number of counts can be substantially higher in sources with complex absorption.

Secondly, the AGN must possess a broad Fe Kα line of sufficient strength relative to the continuum to allow its red wing to be successfully located; usually this corresponds to a line equivalent width of $EW \gtrsim 100$ eV. The first and best-studied broad iron line in an AGN is that of MCG–6-30-15, which was initially discovered in an *ASCA* observation by Tanaka et al. (1995), largely because of its high equivalent width ($EW = 330^{+180}_{-120}$ eV). The strength and breadth of this feature have been confirmed in subsequent *XMM-Newton* and *Suzaku* observations by many authors, most recently Brenneman and Reynolds (2006), Miniutti et al. (2007), and Chiang and Fabian (2011). Not all type 1 AGN have been observed to possess such features, however. Recent surveys of hundreds of AGN with *XMM-Newton* have concluded that broadened Fe Kα lines are only present in $\sim$40% of all bright, nearby type 1 AGN (de le Calle Pérez et al. 2010; Nandra et al. 2007), and some broad iron lines have been ephemeral, appearing and disappearing in the same object observed during different epochs (e.g., NGC 5548, Brenneman et al. 2012).

L. Brenneman, *Measuring the Angular Momentum of Supermassive Black Holes*, SpringerBriefs in Astronomy, DOI 10.1007/978-1-4614-7771-6_3, © Laura Brenneman 2013

Thirdly, the Fe Kα line in question must be *relativistically* broad in order to be used to constrain black hole spin; that is, it must have a measured inner disk edge—assumed to correspond to the ISCO—of $r_{in} \leq 9\,r_g$. Because the measurement of spin is predicated on the assumption that the inner edge of the disk truncates at the ISCO, the traditional approach when employing spectral models for the iron line (and other reflection features) that allow black hole spin to be a free parameter in the fit is to fix $r_{in} = r_{ISCO}$. A valuable first step before applying such a model, however, is to assess the location of the inner edge of the disk by first fitting the iron line with a fixed-spin model such as `diskline` ($a = 0$; Fabian et al. 1989) or `laor` ($a = +0.998$; Laor 1991) and allowing the r_{in} to fit freely. If $r_{in} \lesssim 9\,r_g$, subsequent fitting with a free-spin model will return believable spin constraints.

Though line-only models are useful as a tool to establish an initial starting point in parameter space for spectral fitting, it is important to model the inner disk full reflection spectrum when making spin measurements, not just the broad Fe Kα line. Separate, line-only models can be misleading when attempting robust spin measurements because they do not take into account the curvature of the reflection spectrum over the full energy range of the data, e.g., from the Compton hump and associated iron edge absorption. Only an accurate, holistic modeling approach that treats the continuum, absorption and the entire reflection spectrum at once will yield robust, meaningful constraints on black hole spin.

Taking all these points into consideration, the potential sample size of spin measurements for AGN in the local universe is $\sim$30–40 sources (Miller 2007). Most of these AGN are type 1, lacking significant obscuration by dust and gas along the line of sight to the inner disk. Though it is possible to obtain spin measurements for more heavily absorbed type 2 sources, great care must be taken to properly account for all of the absorption in the system.

3.2   Spectral Models

The reflection spectrum from the inner disk can be self-consistently reproduced by models such as `reflionx` (Ross and Fabian 2005; see Fig. 2.2) or `xillver` (Garcia and Kallman 2010). These models simulate not only the broad Fe Kα line, but all other fluorescent emission species at lower energies, as well as the Compton hump at higher energies. Typical free parameters include the ionization of the disk and its iron abundance (assumed, simplistically, to be constant values), the photon index of the irradiating power-law continuum (usually tied to that of the power-law itself), and the flux or normalization of the reflection spectrum. In order to incorporate the effects of relativity and Doppler shift, this static reflection spectrum must then be convolved with a smearing algorithm which computes the photon trajectories and energies during their propagation from the accretion disk to the observer. Several free-spin smearing algorithms are currently available for use (see Chap. 1). The `kerrconv` algorithm of BR06 is the only one of these models that is currently built into XSPEC, though it limits black hole spin to prograde values only. A more recent improvement is the relconv model of Dauser et al. (2010, 2013), which generalizes the possible spins to incorporate retrograde black holes (see Fig. 2.3).

Isolating the inner disk reflection spectrum from other X-ray spectral signatures is often the greatest challenge in obtaining robust constraints on black hole spin, even in deep ($\gtrsim$100,000 s) observations of bright AGN. Due to the relatively poor spatial resolution of X-ray telescopes, the spectra obtained from the cores of AGN represent a superposition of emission and absorption from several different physical processes within these regions. The principal components of AGN X-ray spectra (other than relativistic reflection) and their physical origins are enumerated below. Most, though not all, of these components are present in any given AGN.

1. **Continuum Emission:** As discussed in Chap. 2, the source of the continuum in X-rays is Compton upscattering of thermal photons from the accretion disk. At present very little is known about the origin, geometry and location of the hot electrons responsible for this scattering, but the timescales of variability (~hours) for the power-law component in AGN spectra suggest that this corona is compact and likely also close to the accretion disk (e.g., Markoff et al. 2005). Magnetic fields almost certainly play a critical role in its formation, perhaps facilitating the heating and/or flaring of certain regions of the electron plasma (Di Matteo 2001). Though this corona is readily approximated by a power-law with a high-energy cutoff (a proxy for the electron temperature), more physical models have been created which include free parameters for optical depth, electron temperature, seed photon temperature, compactness, thermal vs. non-thermal electron population, etc. Three of the most popular of these models are compTT (Titarchuk 1994), its successor, compPS (Poutanen and Svensson 1996), and eqpair (Coppi 1999).

2. **Intrinsic Cold (Neutral) Absorption:** The putative molecular torus of AGN unification schemes (Antonucci 1993; Urry and Padovani 1995) manifests in two forms in X-rays: as the source of the primary neutral absorbing column within the nucleus, and as a scattering medium for the continuum emission, forming the distant reflection signatures discussed in item #4 below. This reservoir of gas is a relatively cold, neutral, optically-thick medium thought to reside on the order of 10^4–$10^5 \, r_g$ from the black hole. Its gas appears to be anisotropically distributed, though its origins are still a topic of active research. The radius of this neutral gas from the black hole, with respect to the broad emission line region (BELR), may vary from object to object, however. For example, NGC 1365 shows evidence for a clumpy absorber at a radius $r \leq 2 \times 10^{15}$ cm from the black hole, well within the BELR, in which the clumps of gas eclipse the inner disk/corona region (e.g., Maiolino et al. 2010; Risaliti et al. 2005a, Brenneman et al. 2013). Recent work suggests that radiation pressure may play a prominent role in forming these collections of neutral gas (Elvis 2012). Whatever its origin, this cold absorbing gas is best modeled with a simple photoelectric absorption component whose low-energy cutoff is determined by the column density of the gas (e.g., phabs, within XSPEC, or tbabs, from Wilms et al. 2000), or a partial-covering variant

of these models with an additional parameter for the covering fraction of the gas over the continuum source (e.g., `pcfabs`, within XSPEC).

3. **Intrinsic Warm (Ionized) Absorption:** Some of the first signatures of ionized absorbing gas within AGN were originally reported by Halpern (1984), though they were not detected commonly in X-rays until the *ASCA* era (e.g., Reynolds 1997). Early CCD resolution could only detect the two most obvious manifestations of these features, the O VII and O VIII absorption edges at 0.74 and 0.87 keV, respectively. But with the advent of grating spectroscopy in the *Chandra* and *XMM-Newton* era, warm absorbers are now known to harbor a rich forest of lines and edges from various species of gas and dust (Lee 2010). The presence of these features can extend up into the Fe K band in some cases where the ionization of the gas is high enough, and are sometimes associated with outflows having speeds of up to $v_{out} \sim 0.4c$ based on their blueshifted absorption features (e.g., Tombesi et al. 2010). Most warm absorbers are now thought to incorporate several "zones" of material, distinct in their kinematic properties as well as their column densities and ionizations, but maintained in pressure balance (e.g., NGC 3783; Krongold et al. 2003). Though they can be modeled with individual absorption lines and edges, the most self-consistent way to model these features over an entire spectrum is with tables produced by spectral synthesis codes (including radiative transfer) such as CLOUDY (Ferland et al. 2013) or XSTAR (Kallman and Bautista 2001).

4. **Distant Reflection:** When irradiated by the X-ray power-law continuum emission, the outer disk/torus produces a reflection spectrum much like that of the inner accretion disk, only without the convolved relativistic effects due to its distance from the black hole (George and Fabian 1991; Matt et al. 1992). The narrow Fe Kα line and Compton hump are the two most prominent features of this distant reflection spectrum; in fact, the narrow Fe Kα line appears to be nearly ubiquitous in Seyfert 1 galaxies (Yaqoob and Padmanabhan 2004). Though the Compton hump can be modeled adequately with a `pexrav` component (Magdziarz and Zdziarski 1995), one must include additional Gaussian emission lines to model the discrete Fe K features. Alternatively, the `pexmon` model (Nandra et al. 2007) includes the Fe Kα, Kβ and Ni Kα lines as well as the Compton shoulder of the Fe Kα line self-consistently with the `pexrav` reflected continuum. There are two important caveats to keep in mind when applying these models, however: (1) `pexrav` and `pexmon` are designed to simulate the spectrum produced from the irradiation of a thin disk rather than a puffy or toroidal structure, so this model may not be an accurate representation of the system; (2) neither model accounts for the contribution of the cold reflected emission at energies $E \lesssim 3$ keV. To consider the contribution from emission below this energy, one can also use a static `reflionx` model to simulate the contribution from the torus across all energies, though this model also assumes a disk geometry. The `MYTorus` model of Murphy and Yaqoob (2009) is a more physical alternative that takes the geometry of the reprocessing medium into account, as well as its low-energy emission, though at present it is unclear how

well the opening angle of the reprocessor can be constrained, and the model does not allow the iron abundance of the reprocessor to vary freely.

5. **Soft Excess Emission:** Some of the first observational evidence for a "soft excess" in X-rays above the power-law continuum was observed by *EXOSAT* in Mrk 841 (Arnaud et al. 1985). Though this emission can be well fit with a blackbody or modified disk blackbody (`diskbb`) component, however, the typical temperature ($\sim$0.2 keV) is too high to be thermal emission from the disk in an AGN, as a rule. Advanced CCD and grating spectra made possible in the *Chandra/XMM-Newton/Suzaku* era have since demonstrated both the ubiquity of this feature in AGN with moderate-to-high accretion rates, and that the typical temperature of this component does not appear to change with source luminosity, posing theoretical problems for many thermal and single-zone Comptonization models (Bianchi et al. 2009; Done et al. 2012). Other possible origins for the soft excess include a contribution from inner disk reflection, photoionized or extended thermal plasma emission (e.g., from a circumnuclear starburst region), scattering of continuum photons, or emission from the base of a jet (see Lohfink et al. 2013a and references therein). Because the soft excess typically manifests as a smooth feature, an adequate fit can be equally well achieved with any (or more than one) of the above modeling components, leading to uncertainty about its physical origin. The best-fit model component(s) of the soft excess also seem to vary between AGN. Effective modeling of the soft excess is critical to constraining the physical parameters of the AGN spectrum, however, because X-ray detectors tend to have higher collecting areas at lower energies where this component dominates (i.e., $\leq$2 keV). The fit statistic will therefore be dominated by this region of the spectrum.

3.3 Time-Averaged vs. Time-Resolved Spectra

An examination of the time-averaged spectrum maximizes the S/N of the observation and allows one to assess which physical components are present in the data. By contrast, time-resolved spectroscopy is critical for identifying and properly modeling the various physical components of a typical AGN system. Many of these components may vary substantially during the observation, and those variations may appear averaged-out and provide misleading information about the source when viewed through the time-averaged spectrum alone.

Whenever possible (i.e., given enough S/N in reasonable time bins, $\sim$10,000 counts per bin), a time-resolved spectral analysis should be undertaken in addition to the time-averaged analysis if the source flux and/or spectral hardness varies substantially during the observation. Data from different time intervals and/or flux states within an observation (or indeed, data of the same source taken during multiple epochs) can be analyzed jointly in order to assess the physical nature and variability of all of the components in a given X-ray spectrum, often yielding a more physical picture of the system whose changes during an observation can be washed out in a time-averaged spectrum.

For example, the power-law and inner disk reflection are expected to vary on timescales as short as hours in a typical AGN if the continuum emission and reflection are centrally concentrated (e.g., Miniutti and Fabian 2004; Uttley et al. 2005). By contrast, distant reflection from the outer disk/torus region typically varies on the order of ∼days–weeks (McHardy et al. 1999), and warm absorbers can show changes in their column densities and/or ionizations on timescales of ∼weeks–months (Krongold et al. 2005). Both of these components take longer to respond to changes in the continuum emission than the inner disk radii due to their relatively large distances from the corona. The soft excess in many AGN is often unrelated to source luminosity and can be constant over years-long timescales (Crummy et al. 2006). By jointly analyzing spectra from different time or flux intervals in a given AGN, one can tie certain model parameters together between intervals if they are not expected to vary during or between observations, while leaving the other model parameters free to vary. Doing so effectively increases the S/N of the data and yields more accurate constraints on slowly- or non-varying parameters of these systems (e.g., black hole spin, disk inclination, iron abundance).

Time-resolved spectroscopy is particularly useful for disentangling the effects of complex absorption from the properties of the continuum and inner disk reflection, because the majority of AGN show evidence for absorption in their spectra. Some researchers have even proposed that, in many AGN with purported broad Fe Kα lines, these apparent reflection features are actually artifacts of improperly modeled absorption. For example, Miller et al. (e.g., 2008, 2009; MTR) argue that the archetypal broad iron line AGN, MCG–6-30-15, actually shows no relativistic inner disk reflection, but instead has five layers (or "zones") of absorbing gas intrinsic to the AGN, covering a wide range in column density, ionization parameter and covering fraction. The superposition of spectral features created by these absorbing structures mimics the appearance of relativistic reflection features. The MTR absorber's incorporation of partial covering, especially, is what allows the model to achieve a goodness-of-fit comparable to that of the relativistic reflection model, which does include complex absorption, but with fewer zones. The debate is ongoing regarding which model is a more plausible physical representation of the system, and a combination of a broad spectral bandpass and time-resolved and/or multi-epoch spectral analysis are needed to definitively address this question (see Chaps. 5 and 7). When viewed holistically in this manner, the relativistic reflection model (e.g., that of BR06, Miniutti et al. 2007, Chiang and Fabian 2011) will vary in ways that have no analog in the MTR absorber model, and vice versa.

In the following sections, I describe the practicalities of using the relativistic reflection model (employing a `reflionx` disk reflection model and either a `kerrconv` or `relconv` smearing algorithm) to measure the spins of the SMBHs in three well-known AGN using *XMM-Newton*, *Chandra* and *Suzaku* observations: MCG–6-30-15, NGC 3783 and Fairall 9.

Chapter 4
Case Studies: MCG–6-30-15, NGC 3783 and Fairall 9

4.1 MCG–6-30-15

The type 1 AGN MCG–6-30-15 ($z = 0.0077$) was the first galaxy in which a broad Fe Kα line was observed, using *ASCA* spectra (Tanaka et al. 1995). This feature still stands as the broadest line of its kind to date, with a red wing extending down to $\sim$3 keV, and a typical strength measured at $\sim$200 eV (Fabian et al. 2002). As such, MCG–6-30-15 is one of the most observed AGN in X-rays, with numerous pointings from *Chandra, XMM-Newton* and *Suzaku* over the past decade in the public archives.[1]

The first measurement of the SMBH spin in MCG–6-30-15 using the technique described above in Chap. 3 was made by BR06 using *XMM-Newton* data from the long 2001 observation first reported by Fabian et al. (2002). BR06 constrained the spin to $a \geq +0.98$ to 90% confidence with a model incorporating a kerrconv smearing kernel acting on a reflionx disk reflection spectrum, a three-zone, dusty warm absorber and a soft excess that was modeled equally well with either a blackbody or compTT component. The limited spectral range of *XMM-Newton* (0.3–12 keV) posed difficulties for modeling the Compton hump, however, since this feature lies mostly outside of the telescope's energy band.

The launch of *Suzaku* in 2005 enabled the 0.3–12 keV energy range to be complemented by simultaneous data up to $\sim$60 keV for bright AGN, using *Suzaku*'s XIS and PIN instruments in tandem. Miniutti et al. (2007) examined the *Suzaku* spectrum of MCG–6-30-15 for the first time, using $\sim$330 ks of data obtained over 2 weeks in January 2006. The flux of MCG–6-30-15 at this time was measured at 4.0×10^{-11} erg cm^{-2} s^{-1}, yielding 1.98×10^6 counts in the XIS detectors and 1.50×10^5 counts in the PIN detector. Miniutti et al. noted the striking similarity between the *Suzaku* and *XMM-Newton* data in the Fe K band (see Fig. 4.1). They then restricted their energy range to $\geq$3 keV in order to avoid most of the spectral

---

[1]NASA/GSFC maintains the High-Energy Astrophysics Science Research Center, or HEASARC: http://heasarc.gsfc.nasa.gov/.

L. Brenneman, *Measuring the Angular Momentum of Supermassive Black Holes*, SpringerBriefs in Astronomy, DOI 10.1007/978-1-4614-7771-6_4, © Laura Brenneman 2013

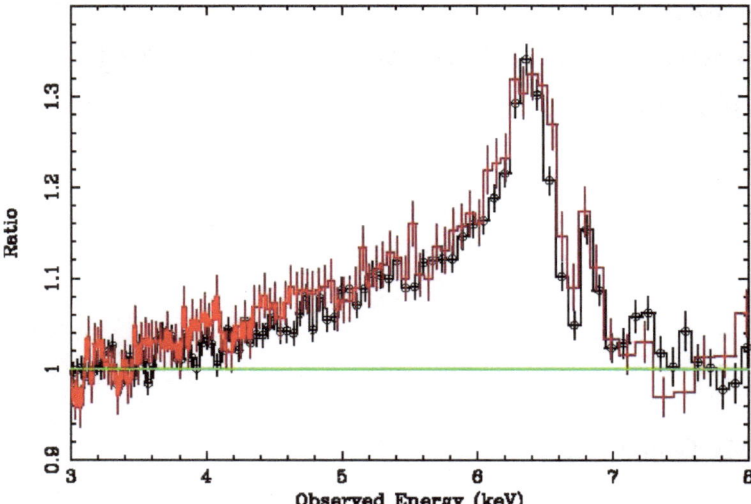

Fig. 4.1 *Suzaku* (*black*, 2006) and *XMM-Newton* (*red*, 2001) data showing the broad Fe Kα line in MCG–6-30-15 ratioed against a power-law continuum. The feature shows little variation in strength and breadth between epochs. *Solid black* and *red lines* are meant to guide the eye and do not represent a model. The *solid green line* shows a data-to-model ratio of unity (Figure is from Miniutti et al. (2007). Reprinted by permission of *Publ. Astr. Soc. Japan*)

complexities associated with the warm absorber and used a model similar to that of BR06 to derive a black hole spin of $a \geq +0.92$.

One of the most comprehensive analyses to date of MCG–6-30-15 was performed by Chiang and Fabian (2011; CF11). These authors re-examined the *XMM-Newton+BeppoSAX* (2001), *Chandra* (2004) and *Suzaku* (2006) data jointly in order to characterize the nature and variability of the warm absorber, continuum and reflection components of the source holistically. The *Suzaku* spectrum is shown in Fig. 4.2 as a ratio to the power-law continuum and Galactic photoabsorbing column in order to illustrate the various residual spectral features present. A very strong Compton hump is visible above 10 keV and rolls over at ∼20 keV. At lower energies, the narrow Fe Kα line is prominent at 6.4 keV along with an absorption line of Fe XXV at ∼6.7 keV and a small Fe Kβ emission line at ∼7 keV. The broad line is quite strong, extending down to ∼3 keV on the red wing. Below this energy, the spectrum takes on a concave shape due to the presence of absorption by gas at lower ionization states. A weak soft excess is evident below ∼0.8 keV.

The goal in modeling the spectrum over the entire energy range is to achieve the best possible statistical fit with the lowest possible number of parameters using a physically self-consistent approach. CF11 assumed that the basic components of the fit are the same as those seen in all other type 1 AGN, as described in Sect. 3.2: power-law continuum, distant and inner disk reflection, complex absorption and a soft excess. No *a priori* constraints were placed on the physical nature of the soft

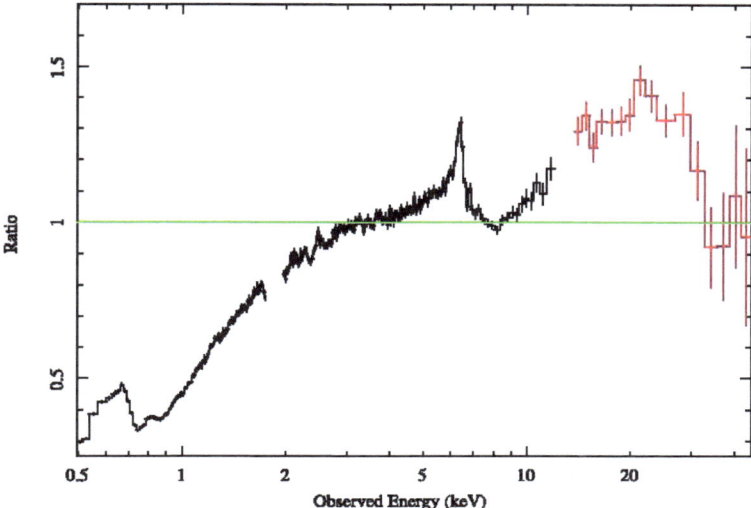

Fig. 4.2 *Suzaku*/XIS (*black*) and PIN (*red*) data from the 2006 observation of MCG–6-30-15, here shown ratioed against a power-law. The *black* and *red solid lines* connect the data points and do not represent a model. The *solid green line* shows a data-to-model ratio of unity (Figure is from Miniutti et al. (2007). Reprinted by permission of *Publ. Astr. Soc. Japan*)

excess emission, as this is still a topic of debate within the community and may vary among AGN. The inner disk was assumed to extend radially from $r_{in} = r_{ISCO}$ to $r_{out} = 400\,r_g$. CF11 modeled the spectra from *Chandra, XMM+BeppoSAX* and *Suzaku* accordingly, beginning with the power-law continuum modified by Galactic photoabsorption and progressively adding new components as warranted by an improvement in the fit statistic.

Absorption modifies an entire spectrum in a multiplicative sense ($\propto e^{-\tau}$, where τ is the optical depth of the absorbing gas), meaning that it can affect the shape of the overall spectrum across a significant fraction of the energy band. As such, the absorption should be addressed early during modeling because it will have a significant effect on the parameters of the continuum (most notably the slope of the power-law). CF11 employ XSTAR tables to model the multi-zone warm absorber in MCG–6-30-15, allowing the column density and ionization to be free parameters in the fit. Primarily informed by the high-resolution *Chandra*/HETG data, the authors find that three zones of ionized absorbing gas intrinsic to the AGN are required to properly model its spectral curvature.

To illustrate that these three absorption components are both necessary and sufficient, CF11 consider difference spectra created by subtracting the spectrum of the low-flux state of the source from that of the high-flux state of the source in each observation (see Fig. 4.3). Although examining a difference spectrum is a common technique used to assess the contribution of additive components (e.g., individual emission lines) to a spectrum at different times, the multiplicative components (e.g.,

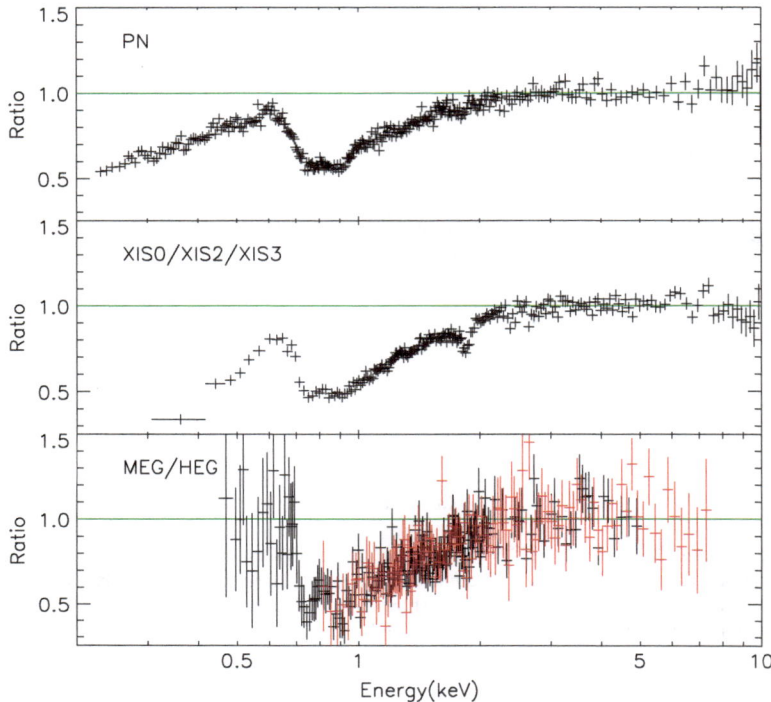

Fig. 4.3 The difference spectra from *top* to *bottom* are extracted from *XMM-Newton*/PN, *Suzaku*/XIS-FI and *Chandra*/HETG, respectively. The absorption structure around 1.8 keV in the *middle panel* is an uncalibrated silicon feature produced by he detector. Note the similarity in spectral shape between the three observations (Figure is from Chiang and Fabian (2011). Reprinted with permission from Oxford University Press)

absorption) also manifest because they cannot be subtracted out. The difference spectra drawn from the three satellites are shown here as a ratio to a simple power-law model modified by Galactic absorption. They are remarkably similar at energies below 3 keV. An obvious drop below 2 keV is seen and goes to nearly the same depth in all difference spectra. This suggests that the low energy spectra can be represented by the same model, and that the warm absorber does not change significantly between the three observations. The 3–10 keV (3–7.5 keV for *Chandra*) difference spectra can be fitted by a simple power-law, implying that the warm absorber causes little curvature above 3 keV and that the (additive) reflection signatures are largely unchanged between epochs.

The warm absorber does not appear to mimic the red wing structure seen in the Fe K band, nor the excess emission seen at higher energies that is commonly attributed to the Compton hump also produced by reflection. This contrasts with the model suggested by Miller et al. (2008, 2009) which incorporates two additional partial-covering clumpy absorbing zones to model the high-energy spectrum, in particular. One of these zones mimicked the shape of broadened Fe Kα line, and the other partially covered the continuum in order to explain the hard excess. The main difference between the absorption-only model and the reflection-dominated model is that the former has no distortions due to relativistic effects.

Miller et al. (2009) have claimed that the reflection model fails to interpret the hard excess. However, CF11 showed that the hard excess can be simply explained by a relativistically blurred reflection component of super-solar iron abundance *without* any additional, partial-covering absorption. However, it should be noted that even though the Miller et al. model does not address the difference spectra of MCG–6-30-15, both the reflection and absorption-only models do provide adequate fits to the spectrum and variability seen in MCG–6-30-15 in these data.

Within the framework of the reflection model, both distant and inner disk reflection components must be included in order to accurately model the spectrum of MCG–6-30-15. CF11 parametrize these features with two reflionx models, convolving the inner disk component with a kdblur smearing kernel analogous to the laor relativistic line profile (i.e., SMBH spin fixed at $a = +0.998$) while leaving the distant reflionx component unsmeared. Though it is not possible to formally constrain black hole spin using the approach these authors have chosen, the measurement of the inner disk radius obtained via this method can provide some insight into the magnitude and direction of the black hole's angular momentum. The best fit obtained is $\chi^2/\nu = 5,059/3,809\,(1.33)$ for *XMM+BeppoSAX*, $\chi^2/\nu = 2,418/2,139\,(1.13)$ for *Chandra* and $\chi^2/\nu = 1,685/1,576\,(1.07)$ for *Suzaku*. In each case, the good quality of the fit is a strong indication that a rapidly-spinning, prograde black hole resides in MCG–6-30-15. This is confirmed by the constraints placed on the inner edge of the disk, particularly in the *XMM-Newton* observation, which has the highest S/N in the Fe K band: $r_{in} \le 1.7\,r_g$. This equates to a spin of $a \ge +0.97$, entirely consistent with the results of BR06 and Miniutti et al. (2007). The reflectors are both approximately neutral, with a measured iron abundance of Fe/solar $= 1.7^{+0.2}_{-0.1}$ and a disk inclination angle of $i = (38^{+3}_{-2})°$ to the line of sight. The soft excess is represented adequately by the inner disk reflector in the CF11 model, so no additional spectral component is required. The best-fitting model to all three datasets is shown in Fig. 4.4, while the best-fitting model components are shown in Fig. 4.5.

Even if we assume that the reflection-dominated model is the most physically realistic explanation for the spectrum and variability of MCG–6-30-15, controversy over the derived reflection parameters remains. Patrick et al. (2011; hereafter P11) employ a very similar spectral model to CF11 with a three-zone warm absorber, power-law continuum and both distant and inner disk reflection, yet their measured spin parameter is $a = 0.61^{+0.15}_{-0.17}$, more than a 2σ off the value measured by BR06 and Miniutti et al. (2007). However, P11 also model the soft excess with a compTT component that is assumed from the start of their modeling, rather than fitted as a remaining residual after the continuum, absorption and reflection have been accounted for. This compTT component has a modest temperature, optical depth and flux, in keeping with the modest strength of the soft excess in this source ($kT = 3.9\,keV$, $\tau = 0.8$, $F_{0.6-10} = 7.2 \times 10^{-12}\,erg\,cm^{-2}\,s^{-1}$).

Fig. 4.4 The best-fit model of CF11 applied to data from *Chandra*/HETG ±1 order MEG (*top*; *black points* show +1, *red* show −1), *XMM-Newton*+*BeppoSAX* (*middle*; *green/blue points* show *XMM-Newton*/RGS, *black* show *XMM-Newton*/PN, *red* show *BeppoSAX*/PDS) and *Suzaku*/XIS+PIN (*bottom*; *red points* show the XIS-BI, *black* show the co-added XIS-FI and *green* show the PIN). Models fit to each dataset are shown as *solid lines* in the *top panels*. The *bottom panels* on each plot show the data-to-model ratio, where the *solid green line* indicates a theoretical perfect fit (Figure is from Chiang and Fabian (2011). Reprinted with permission from Oxford University Press)

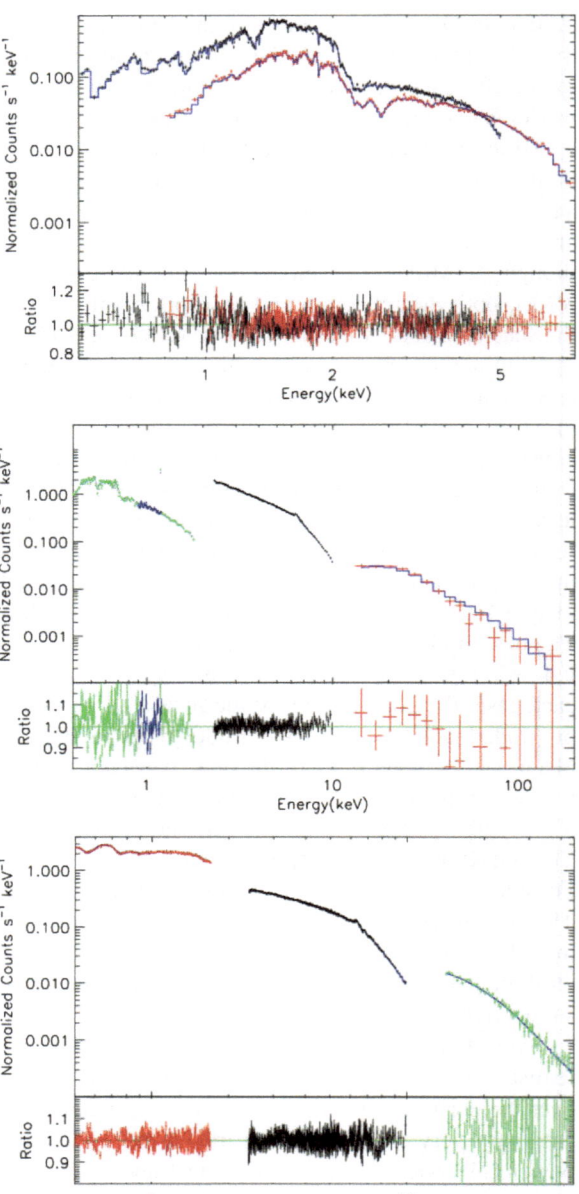

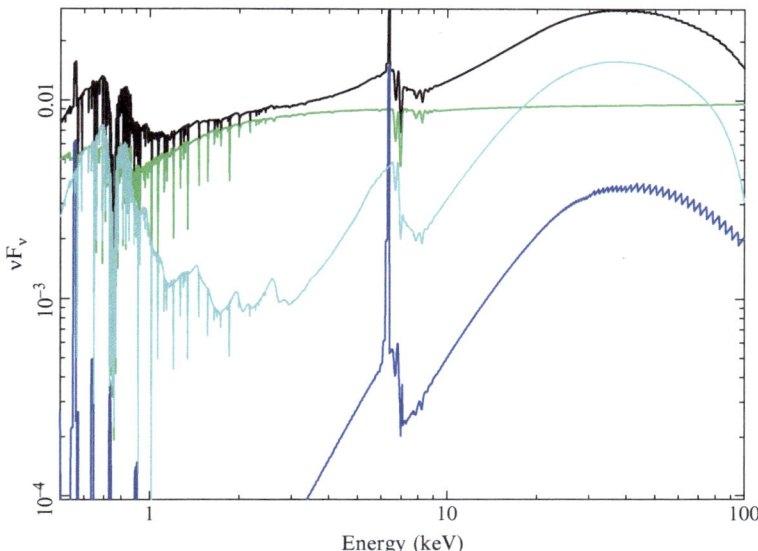

Fig. 4.5 The best-fitting model components for the CF11 model as applied to the 2006 *Suzaku* data for MCG–6-30-15. The total model is depicted in *black*, the power-law in *green*, the distant reflector in *dark blue* and the inner disk reflector in *light blue*. The three-zone warm absorber modifies all spectral components. Figure is created based on the best-fit model from Chiang and Fabian (2011)

Another, and perhaps more critical difference of the P11 model from that of CF11 is in the construction of the warm absorber tables. CF11 use iterative fitting to determine the turbulent velocities of the three warm absorption zones: $v_{\text{turb}} = 500\,\text{km}\,\text{s}^{-1}$ for two zones, $v_{\text{turb}} = 100\,\text{km}\,\text{s}^{-1}$ for the remaining zone. P11, by contrast, keep the turbulent velocity of the two low-ionization zones at $v_{\text{turb}} = 200\,\text{km}\,\text{s}^{-1}$ and fix that of the high-ionization zone to $v_{\text{turb}} = 1{,}000\,\text{km}\,\text{s}^{-1}$. Additionally, CF11 allow the iron abundance of the warm absorber to vary (tying it to that of the distant and inner disk reflector), whereas P11 fix the iron abundance at the solar value. These two differences have significant effects on the appearance of the warm absorber, giving it broader spectral lines and requiring higher column densities to model iron features than would be required if super-solar iron abundance were allowed (as found in CF11, using *Chandra*/HETG data). These differences are likely responsible for the different column densities and ionizations measured for the warm absorber zones by the two groups. Finally, P11 employ an additional neutral, high column, partial-covering absorber to account for a hard excess above 10 keV above the dual reflector model. This absorber has $N_{\text{H}} = 3.4 \times 10^{24}\,\text{cm}^{-2}$ and $f_{\text{cov}} = 50\%$. No such additional absorber is needed in the CF11 analysis; the hard excess is well accounted for by the super-solar iron abundance allowed in the CF11 absorber tables. The low-ionization absorber of P11 possesses much more curvature up into the Fe K band, while the partial-covering absorber takes up the hard excess emission that would otherwise be modeled by the Compton hump in the CF11 model. This combination results in the lower spin derived by P11.

Disregarding the ongoing debate about which model is a more physical representation of the system, the fact remains that a good statistical fit to the spectral data can be achieved by the reflection-dominated model of BR06 and CF11, the absorption-only model of Miller et al. (2008, 2009) and the "hybrid" model of P11 that combines features of both. Breaking the degeneracy between these models will require high S/N over a broad bandpass in X-rays, and good spectral resolution over the ~0.5–10 keV range, especially in the Fe K band (i.e., ~100 eV resolution at 6 keV). These capabilities will enable the continuum, absorption, reflection and any remaining soft or hard excess emission to be accurately and simultaneously characterized based on their discrete and broad-band features.

The recently launched *NuSTAR* telescope (Harrison et al. 2013) will provide the best S/N above 10 keV ever achieved owing to its large collecting area, its unique focusing optics in this energy range, and its low background. These capabilities will allow the differences between the reflection-dominated and absorption-only models at higher energies to be constrained by the quality of the data, enabling the correct model to be identified. In a 150 ks simulation of an observation of MCG–6-30-15, *NuSTAR* conclusively breaks the degeneracy between the two models, whereas the *Suzaku*/PIN instrument does not due to its higher background and lower collecting area (see Fig. 4.6). When used in tandem with instruments such as *XMM-Newton* or *Suzaku*, *NuSTAR* data will also enable the most precise, accurate constraints on black hole spin in AGN to be obtained by isolating the reflection signatures from the other components in the spectrum more reliably than ever before (see Fig. 4.7). The launch of *Astro-H* (Takahashi et al. 2010) in 2015 will further augment this science by introducing the superior resolution of micro-calorimetry into the broad-band spectrum, enabling the discrete features of the warm absorber, soft excess and Fe K band emission lines to be conclusively identified and modeled. The contributions of *NuSTAR* and *Astro-H* to black hole spin science will be further discussed in Chap. 6.

4.2 NGC 3783

The type 1 AGN NGC 3783 ($z = 0.0097$) was the subject of a 210 ks *Suzaku* observation in 2009 as part of the *Suzaku* AGN Spin Survey Key Project (PI: C. Reynolds, lead co-I: L. Brenneman). The source was observed with an average flux of $F_X = 6.04 \times 10^{-11} \, \mathrm{erg \, cm^{-2} \, s^{-1}}$ from 2 to 10 keV during the observation, yielding a total of ~940,000 photon counts over this energy range in the XIS instruments the PIN instrument from 14 to 45 keV, after background subtraction. The results are reported in Brenneman et al. (2011) (hereafter B11).

The spectrum ratioed against the power-law continuum is shown in Fig. 4.8. The Compton hump is readily apparent at energies ≥ 10 keV, though its curvature is relatively subtle compared with more prominent features of its kind (e.g., in MCG–6-30-15). The 6–7 keV energy range of the spectrum is dominated by narrow and

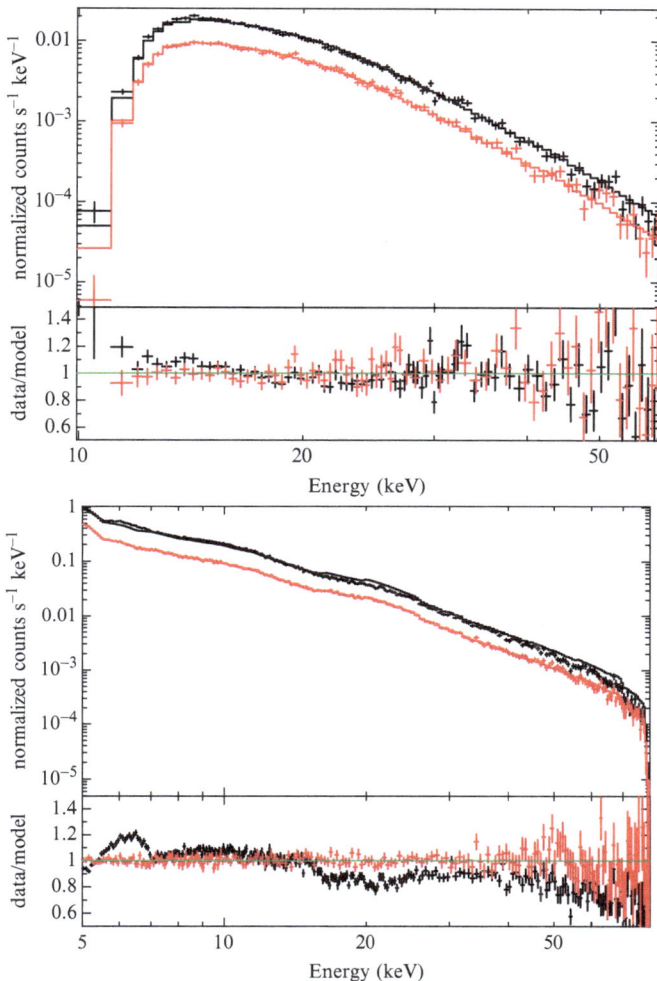

Fig. 4.6 *Top*: The reflection-dominated model of Brenneman and Reynolds (2006) (*black points*) and the absorption-only model of Miller et al. (2008; *red points*) for MCG–6-30-15 simulated with the *Suzaku*/PIN response for 150 ks. Data points are offset in amplitude for clarity. The *black line* in the *top panel* represents the absorption model applied to the simulated reflection data while the *red line* shows the absorption model applied to the absorption data. The data-to-model ratio in the *bottom panel* shows that the two models are not noticeably different to the eye; the residuals from both simulated datasets are clustered about the *green line* (representing a perfect fit) in approximately the same way. $\chi^2/\nu = 1.48$ for the joint fit. *Bottom*: The same plot, but with the two models simulated through the *NuSTAR* response for 150 ks. In this case, applying the absorption model to both simulated datasets results in a much more divergent fit above 10 keV both to the eye and statistically: $\chi^2/\nu = 3.55$ for the joint fit. This demonstrates that NuSTAR can conclusively differentiate between the two models, whereas the Suzaku/PIN instrument cannot

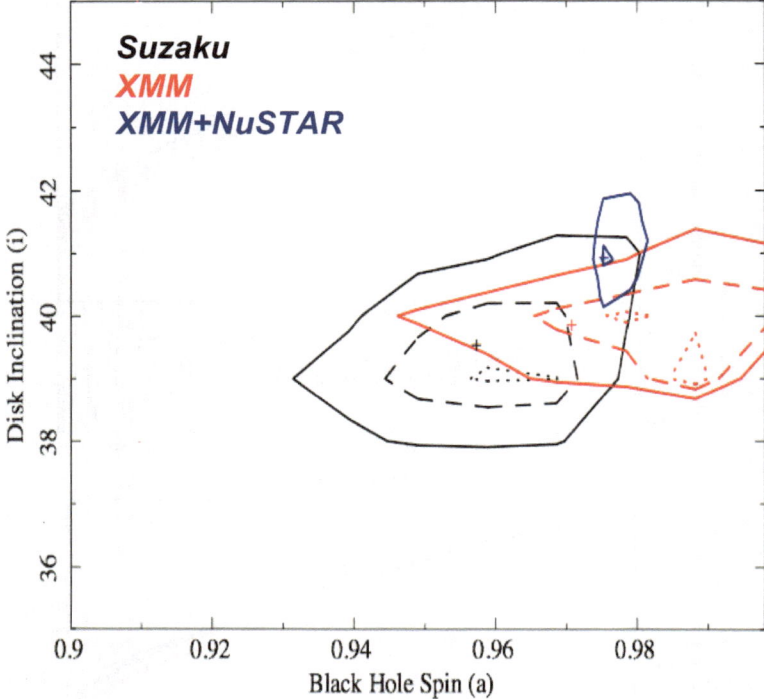

Fig. 4.7 Contour plot of the constraints placed on black hole spin and disk inclination angle in MCG–6-30-15 with *Suzaku* (*black*), *XMM-Newton* (*red*) and *XMM+NuSTAR* (*blue*). *Solid lines* show the 67% confidence region, *dashed lines* show 90% confidence, and *dotted lines* show 99% confidence. These contours are derived from simulations of the BR06 model run through the response of each detector, then refit iteratively with the model in order to assess the confidence intervals for each parameter. Simultaneous *NuSTAR/XMM-Newton* or *NuSTAR/Suzaku* data will improve the precision of the spin constraint by a factor of ∼10 for MCG–6-30-15, while also improving the accuracy of the measurement

broad Fe K features, including a narrow Fe Kα emission line at 6.4 keV and a blend of Fe Kβ and Fe XXVI emission at ∼7 keV. The broad Fe Kα line manifests as an elongated, asymmetrical tail extending redwards of the narrow Fe Kα line to ∼4–5 keV. The Fe K region can be seen in more detail in Figs. 4.9 and 4.10. At energies below ∼3 keV the spectrum becomes concave due to the presence of complex, ionized absorbing gas within the nucleus of the galaxy; the gas is ionized enough that some contribution from this absorber is seen at ∼6.7 keV in an Fe XXV absorption line. There is a rollover back to a convex shape below ∼1 keV, however, where the soft excess emission dominates.

Brenneman et al. (2011) began their model fitting with the continuum power-law and Galactic photoabsorption, then progressively added various model components to represent the residual spectral features, provided that these added

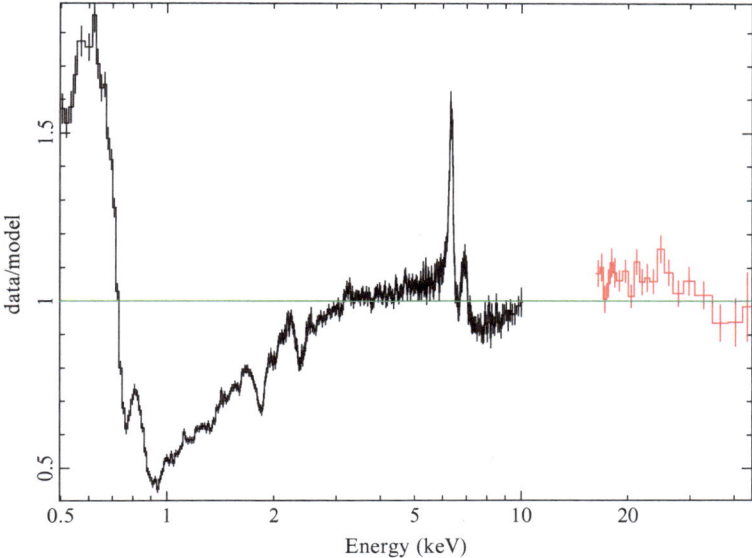

Fig. 4.8 *Suzaku* XIS-FI (front-illuminated; *black crosses*) and PIN (*red crosses*) data from the 210 ks observation of NGC 3783 in 2009, ratioed against a simple power-law model for the continuum (fit over 2–4.5 and 7.2–10 keV) affected by Galactic photoabsorption. *Black* and *red solid lines* connect the data points and do not represent a model. The *green line* depicts a data-to-model ratio of unity. Data from the XIS back-illuminated CCD (XIS-BI) are not shown for clarity (Figure is from Brenneman et al. (2011). Reproduced by permission of the AAS)

Fig. 4.9 A zoomed-in view of the Fe K region in the 2009 *Suzaku* observation of NGC 3783, ratioed against a simple power-law continuum. Note the prominent narrow Fe Kα emission line at 6.4 keV and the blend of Fe Kβ and Fe XXVI at ∼7 keV. XIS-FI is in *black*, XIS-BI in *red* (Figure is from Brenneman et al. (2011). Reproduced by permission of the AAS)

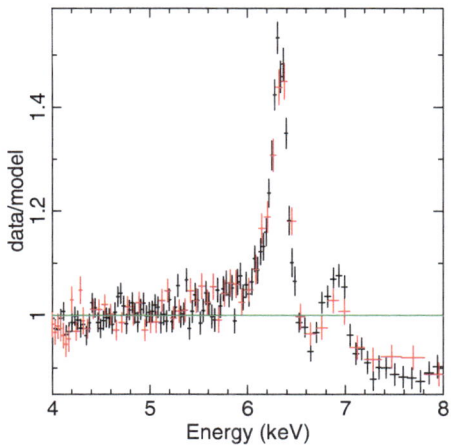

components improved the fit statistically, according to the F-test. A 2001 *Chandra*/HETG observation of this AGN was used to inform the modeling of the warm absorber, since *Suzaku*'s CCDs lack the resolution of the *Chandra* gratings. Though warm absorbers in AGN tend to vary on timescales of ∼weeks–months

Fig. 4.10 The broad Fe Kα
line at 6.4 keV becomes more
obvious when the two more
prominent narrow emission
lines are modeled out, in
addition to the power-law
continuum (Figure is from
Brenneman et al. (2011).
Reproduced by permission of
the AAS)

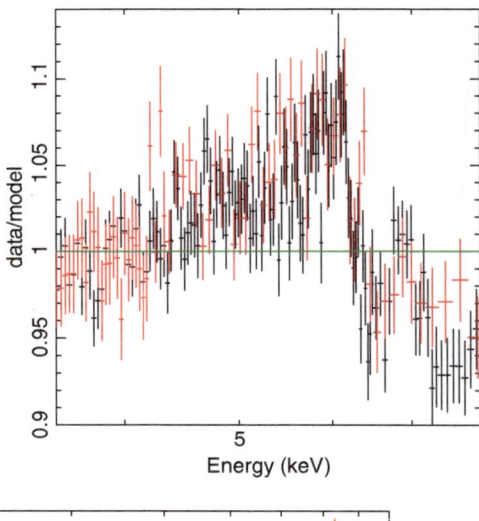

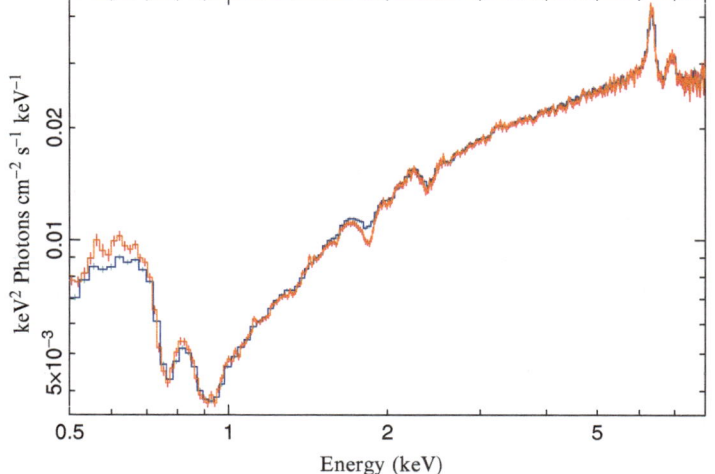

Fig. 4.11 Plot of the best-fitting model from the 2001 *Chandra*/HETG data (*red*) on *top* of the
2009 *Suzaku*/XIS-FI data (*blue*) of NGC 3783. No refitting has been performed, and the *Chandra*
model has been folded through the *Suzaku*/XIS response. Note the excellent agreement between
the 2009 data and 2001 model, indicating that the warm absorber is in a very similar state in these
two observations (Figure is from Brenneman et al. (2011). Reproduced by permission of the AAS)

(Krongold et al. 2005), the 2001 *Chandra* data were a surprisingly good match for
the 2009 *Suzaku* data in terms of absorber appearance, enabling their use in this
capacity (see Fig. 4.11).

Brenneman et al. (2011) used the models and methods described above in Chap. 3
to fit the 0.7–45 keV *Suzaku* spectrum of NGC 3783 with a statistical quality of
$\chi^2/\nu = 917/664\,(1.38)$. Most of the residuals in the best-fit model manifested

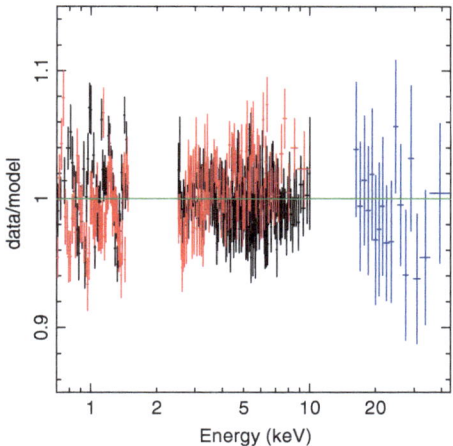

Fig. 4.12 Data/model ratio of the best-fitting model for the 2009 *Suzaku* observation of NGC 3783. XIS-FI data are in *black*, XIS-BI data are in *red* and PIN data are in *blue*. The *green line* represents a data/model ratio of unity. Energies from 1.5 to 2.5 keV and 10 to 16 keV are ignored due to calibration uncertainties (Figure is from Brenneman et al. (2011). Reproduced by permission of the AAS)

below $\sim$3 keV in the region dominated by the warm absorber and soft excess, as is typical for type 1 AGN. Because the S/N of the XIS detectors is highest at lower energies due to the higher collecting area there, small residuals in the spectral modeling of this region can have an exaggerated effect on the overall goodness-of-fit. Excluding energies below 3 keV in the fit, B11 achieved $\chi^2/\nu = 499/527$ (0.95). No significant residuals remained. See Figs. 4.12 and 4.13 for the best-fit data/model ratio and relative contributions of the various model components. The best-fit parameters of the black hole/inner disk system included a spin of $a \geq +0.98$, a disk inclination angle of $i = 22^{+3°}_{-8}$ to the line of sight, a disk iron abundance of Fe/solar $= 3.7 \pm 0.9$ and an ionization of $\xi \leq 8\,\mathrm{erg\,cm\,s^{-1}}$ (errors are quoted at 90% confidence for one interesting parameter). These parameters remained consistent, within errors, when energies $\leq$3 keV were ignored in the fit, negating the importance of the soft excess emission in driving the fit to these parameter values.

The results of B11 were corroborated by Reis et al. (2012), who examined the temporal and spectral variability of NGC 3783 within the *Suzaku* observation, and by Reynolds et al. (2012, hereafter R12), who re-examined the time-averaged data using a Markov Chain Monte Carlo (MCMC) analysis to more closely probe the total available parameter space. These authors especially noted the robustness of the rapid black hole spin and super-solar iron abundance found by B11 (see Fig. 4.14). The variability analysis of Reis et al. revealed that the spectrum is principally composed of two elements: a variable soft component and a quasi-constant hard component, similar in nature to that expected from reflection arising from the inner parts of an accretion disk. Further, difference spectra between different flux states during the observation are all well fit by a simple power-law, suggesting that the well-known warm absorber in this source is not variable during the observation and that the variability is due to changes in the power-law continuum flux. An excess of flux appears at energies $\geq$10 keV in the later stages of the observation. This excess

Fig. 4.13 The relative contributions of the various model components for the B11 best-fit to NGC 3783. The *black line* represents the total model, the *green* shows the power-law continuum, *red* shows the blackbody soft excess, *magenta* shows the component of scattered emission, *dark blue* shows the distant reflection and *light blue* shows the inner disk reflection (Figure is from Brenneman et al. (2011). Reproduced by permission of the AAS)

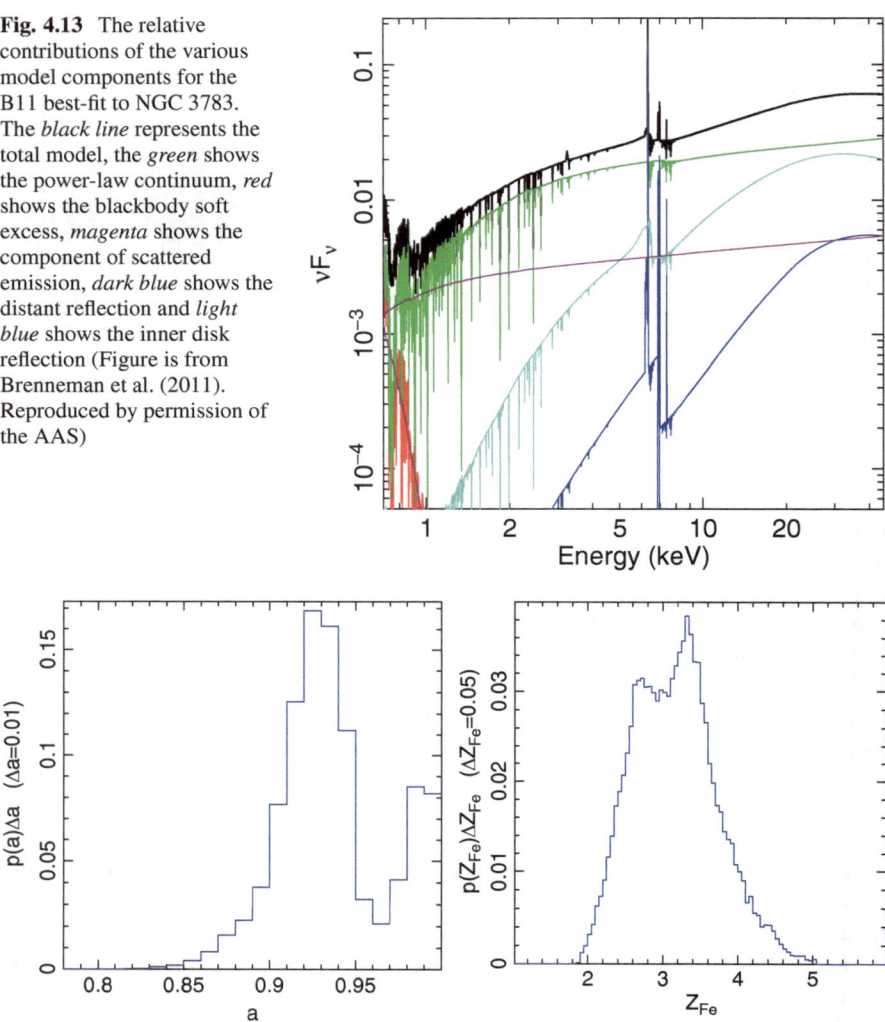

Fig. 4.14 *Left*: The probability distribution of black hole spin for NGC 3783, based on an MCMC analysis (R12). *Right*: The probability distribution for iron abundance in the inner disk, also from the MCMC analysis of R12. Note the preference for high spin and iron abundance. There is a positive correlation between these parameters; see R12 for a more complete discussion (Figure is from Reynolds et al. (2012). Reproduced by permission of the AAS)

was shown to vary with time but not source flux, and can be effectively accounted for by changes in the reflection strength and/or ionization of the inner disk during the observation.

However, P11 analyzed the same data separately and reached a strikingly different conclusion regarding the spin of the black hole in NGC 3783: $a \leq 0.31$.

Fig. 4.15 The relative contributions of the various model components for the P11 best-fit to NGC 3783. The *black line* represents the total model, the *red line* shows the power-law continuum, *magenta* shows the compTT soft excess, *dark blue* shows the distant reflection and *light blue* shows the inner disk reflection. Photoionized emission lines are in *orange* (Figure is created based on the best-fit "dual reflector" model presented in Patrick et al. (2011))

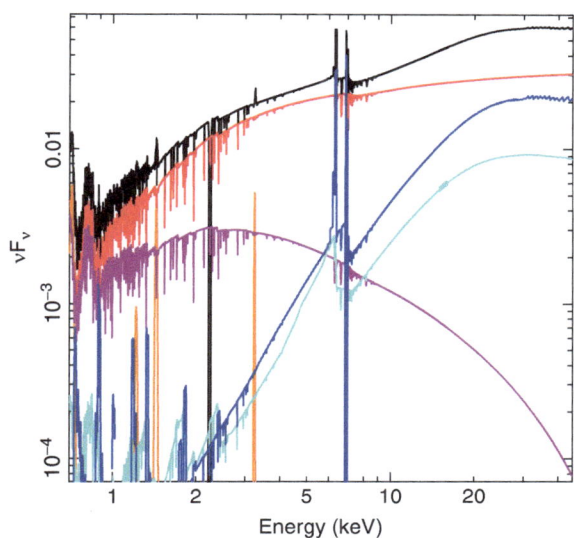

This discrepancy illustrates the importance of assumptions and modeling choices in influencing the derived black hole spin and other physical properties of the black hole/disk system. P11 made three critical assumptions that differed from B11: (1) that the iron abundance of the inner disk is fixed to the solar value; (2) that the warm absorber has a high-turbulence ($v_{turb} = 1,000$ km/s), high-ionization ($\xi \sim 7,400\,\mathrm{erg\,cm\,s^{-1}}$) component not detected by B11; (3) that the soft excess originates entirely through Comptonization, with the Comptonizing medium at a temperature of $kT \geq 9.5\,\mathrm{keV}$ and an optical depth of $\tau = 1.9 \pm 0.1$.

Reynolds et al. (2012) demonstrated that fixing the iron abundance at the solar value significantly worsens the global goodness-of-fit in NGC 3783 when compared with allowing the iron abundance of the inner disk to fit freely ($\Delta\chi^2 = +36$). B11 found no need to include a high-turbulence component in their fit to the *Suzaku* data, and noted no evidence for such a component in the higher-resolution 2001 *Chandra*/HETG data. Finally, R12 note that there is no physical reason to assume that the soft excess originates entirely from Comptonization processes, as other processes within the AGN might contribute (e.g., photoionized emission, scattering, thermal emission). R12 attempted several different model fits to the soft excess and found not only a much smaller contribution to the overall model for the soft excess component than P11, but also no statistical difference between fits using different models (e.g., blackbody vs. compTT). It should be noted, however, that modeling the soft excess with a Comptonization component of high temperature, high optical depth and high flux, as P11 have done, requires the compTT component to possess significant curvature up into the Fe K band, reducing the need for the relativistic reflector to account for this same curvature seen in the data and thereby eliminating the requirement of high black hole spin. To illustrate this, see Fig. 4.15 for a plot of the relative importance of the best-fit model components in the P11 analysis, as

Fig. 4.16 Probability density on the (Z_{Fe}, a)-plane showing a positive correlation. Contour levels are shown at p(Z_{Fe}, a) $= 1, 3.3, 10.0$ (*blue, red, black*; i.e., inner, middle, outer lines), defined such that the probability of being in the range $Z_{Fe} \rightarrow Z_{Fe} + \delta Z_{Fe}$ and $a \rightarrow a + \delta a$ is p(Z_{Fe}, a)$\delta Z_{Fe}\delta a$ (Figure is from Reynolds et al. (2012). Reproduced by permission of the AAS)

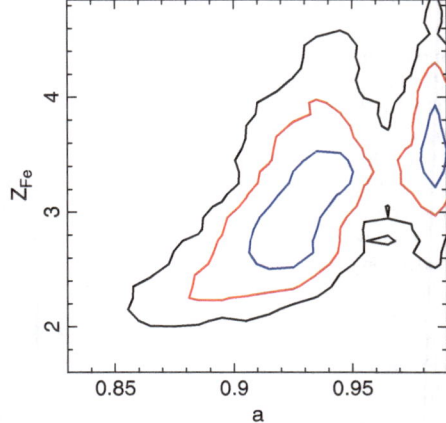

compared with Fig. 4.13 for B11. Clearly, different modeling approaches can lead to vastly different conclusions regarding black hole spin and careful consideration should be given to the models used and to their allowed parameter ranges.

Degeneracies between model parameters can also be a factor and should be carefully considered. For example, R12 discuss the positive correlation between black hole spin and iron abundance found through their MCMC analysis of NGC 3783 (see Fig. 4.16). Both have high values as preferred in the best-fit model, and because increasing the amount of iron in the disk will increase the strength of the reflection features, a rapid spin is required in order to smooth those features out enough to produce an adequate fit to the spectrum. R12 note a worsening of the fit when a fixed, solar iron abundance is adopted, however, lending credence to the super-solar abundance measured.

One possible explanation for the overabundance of iron detected in NGC 3783 and other AGN (e.g., NGC 1365, Walton et al. 2010; 1H0707-495, Zoghbi et al. 2010) is radiative levitation. Previously discussed in the context of surface abundances of white dwarfs (Chayer et al. 1995; Seaton 1996; Wassermann et al. 2010), R12 applied the concept to AGN accretion disks for the first time. Because disks are radiation-pressure dominated in their central regions and can also possess fairly low ionization states of iron (Fe XVII and below) that have populated L- and M-shells, the radiation pressure acting on these heavy metal ions can be much stronger than that acting on the surrounding, fully-ionized plasma. This outward force can also greatly exceed that of gravity, causing the preferential upward drift of iron ions to the disk surface, resulting in an enhancement of iron relative to other elements in the disk atmosphere.

4.3 Fairall 9

As we have seen in the cases of MCG–6-30-15 and NGC 3783, spectral complexities like ionized absorption intrinsic to the AGN can confuse our interpretation of the spectrum. The presence of such components can significantly affect the derived physical parameters of the system, such as black hole spin. It would therefore be ideal to fit relativistic reflection models to a cleaner AGN system without warm absorption as a kind of control case.

There exists a small sample of type 1 AGN, known as "bare" Seyferts, which seem to lack any observable signatures of significant intrinsic absorption in X-rays. While most of these objects do display a soft excess, usually the flux of this component is substantially smaller than that seen in NGC 3783, so the exact model used to parametrize it will have a negligible effect on the spectrum in the Fe K band and will not compromise the spin measured from the broad Fe Kα line and Compton hump.

Fairall 9 is one such bright, nearby ($z = 0.047$), "bare" Seyfert with over 160 ks of data in the *XMM-Newton* archive and nearly 400 ks in the *Suzaku* archive. Though a spectral analysis of the *XMM-Newton* data incorporating relativistic reflection features was reported in Brenneman and Reynolds (2009), the first reported spin measurement for this AGN was published by Schmoll et al. (2009) using a 167 ks *Suzaku* observation from 2007. The source had a flux of $F_X = 1.5 \times 10^{-11}\,\mathrm{erg\,cm^{-2}\,s^{-1}}$ over 2–10 keV at that time and was best fit using a power-law continuum, distant reflection modeled with a `pexrav` and narrow Gaussians for the cores of the Fe Kα and Kβ lines, and ionized inner disk reflection using a `kerrconv` smearing algorithm convolved with `reflionx`. The spin measured with this model was $a = +0.65^{+0.05}_{-0.05}$, significantly less (>6$\sigma$) than the high spin values measured for MCG–6-30-15 and NGC 3783, and perhaps indicative of a different galaxy and SMBH evolution history in Fairall 9 than for the other two AGN considered thus far.

In 2010, a deep *XMM-Newton* observation of Fairall 9 was obtained (130 ks), and from these data a weak spin constraint was established using the `kerrconv` model convolved with `reflionx`: $a = +0.39^{+0.48}_{-0.30}$ (Emmanoulopoulos et al. 2011). Although formally consistent with the Schmoll et al. (2009) result within errors, the disk inclination angle derived by these authors clashed worryingly with that of Schmoll et al.: $i = (64^{+7}_{-9})°$ vs. $i = 44 \pm 1°$.

A 229 ks *Suzaku* observation of Fairall 9 was obtained in 2009 via the *Suzaku* AGN Spin Survey Project, and all four *XMM-Newton* and *Suzaku* pointings have recently been analyzed jointly in Lohfink et al. (2012; hereafter L12). Both *Suzaku* pointings are also discussed in P11. By considering all four epochs of data, L12 note that the source has an average flux consistent with that of the 2007 *Suzaku* observation, but that the source varies in flux by a factor of ~2 over the 2–10 keV band during the 2010 *Suzaku* observation. In spite of the flux variation, the spectral shape remains very similar, with the power-law and distant reflector evident, along with a broad Fe Kα line and a noticeable Compton reflection hump above 10 keV.

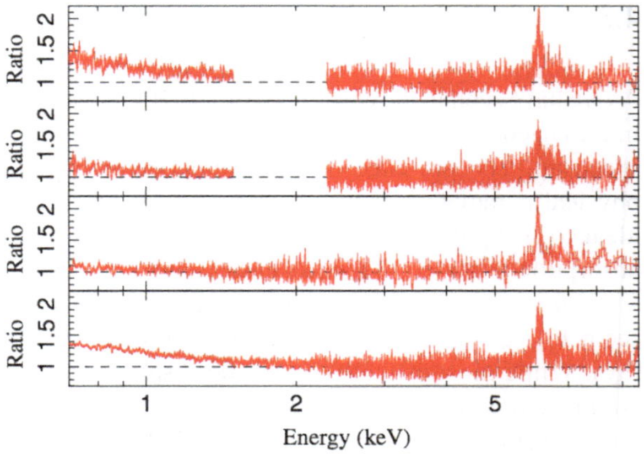

Fig. 4.17 Data-to-model ratios of the Fairall 9 spectra to a simple power-law continuum modified by Galactic photoabsorption. From *top* to *bottom*, the datasets represented are *Suzaku* 2007, *Suzaku* 2009, *XMM-Newton* 2009 and *XMM-Newton* 2000 (Figure is from Lohfink et al. (2012). Reproduced by permission of the AAS)

Fig. 4.18 Unfolded spectrum (*points*) of the 2007 *Suzaku* data in the Fe K band (using a diagonal response), fitted with the best-fitting reflection-only model of L12 (*lines*). The data and best-fit models from all three XIS detectors are shown (Figure is from Lohfink et al. (2012). Reproduced by permission of the AAS)

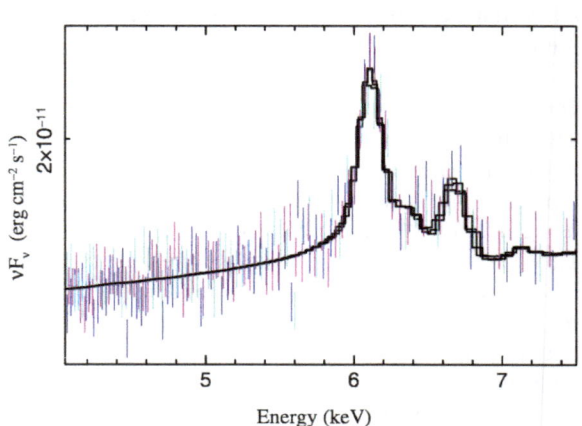

While most of the variation in the flux is evidently due to changes in the power-law strength, a variable soft excess is also visible below 2 keV (see Fig. 4.17). L12 also note the presence of ionized emission lines of Fe XXV and Fe XXVI in the 2009 *Suzaku* spectrum, which are reported in the *XMM-Newton* observations (Brenneman and Reynolds 2009) but not robustly seen in the 2007 *Suzaku* pointing, according to Schmoll et al. (2009). L12 do report these features in the 2007 data, however; see Fig. 4.18 for a close-up look at the Fe K region of this observation.

The best-fitting model obtained by L12 to the four datasets for Fairall 9, fit jointly, requires the standard power-law continuum and near-constant distant reflection (here modeled with a pexmon component), plus reflection from an inner accretion disk with sub-solar iron abundance Fe/solar = 0.67 ± 0.08, ranging in ionization from $\xi = 6^{+3}_{-4}$ erg cm s^{-1} (2007 *Suzaku*) to $\xi = 1{,}739^{+1{,}143}_{-509}$ erg cm s^{-1} (2009 *Suzaku*). The inclination angle of the disk is measured at $i = (37^{+4}_{-2})°$. Additionally, a photoionized plasma is required to explain the ionized iron emission lines seen in the spectrum; this plasma has a loosely constrained ionization of $\xi \sim (0.02–10) \times$

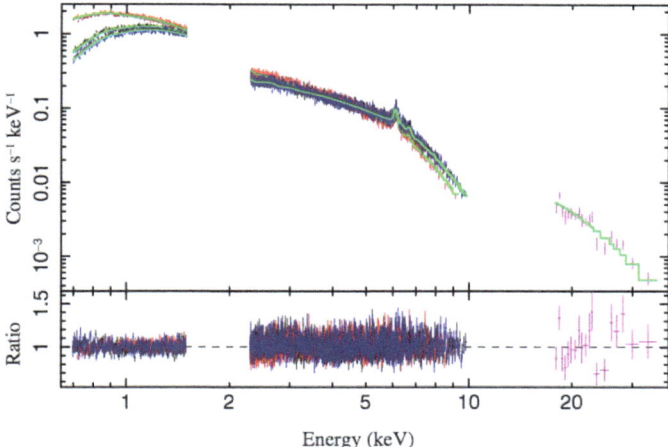

Fig. 4.19 The 2007 *Suzaku* data fit with the reflection-only model of L12. XIS 0 data are the *black points*, XIS 1 data are in *red*, XIS 3 are in *blue* and PIN data are in *magenta*. The model is the *solid green line*. The *lower panel* depicts the data-to-model ratio, which is centered closely about unity (Figure is from Lohfink et al. (2012). Reproduced by permission of the AAS)

$10^6 \, \mathrm{erg \, cm \, s^{-1}}$. The black hole spin is measured at $a = +0.71^{+0.08}_{-0.09}$, consistent with that determined by Schmoll et al. (2009) from the 2007 *Suzaku* data alone. This model yields an excellent goodness-of-fit, with $\chi^2/\nu = 5{,}544/5{,}276 \, (1.06)$ (see Fig. 4.19).

There is still some controversy about the most physical model to use for Fairall 9, however. P11 consider the *Suzaku* data from this source with their dual reflector model, using a `reflionx` component for both the distant and relativistic reflection. In so doing, they find that they must also include a neutral absorber intrinsic to the source ($N_H = 4 \times 10^{23} \, \mathrm{cm^{-2}}$) in order to remove the contribution of the distant reflector from the soft excess emission and fit it adequately with their `compTT` component. These authors also detect the ionized emission from Fe XXV and Fe XXVI, choosing to parametrize the lines with individual Gaussians rather than a photoionized emission model as per L12. There is no clear evidence to either support or disprove the presence of the neutral absorber, which is not reported in any other work on Fairall 9, but it is not necessary to include this component in order to achieve a good fit: the $\chi^2/\nu = 929/881 \, (1.05)$ of P11 is comparable to that of L12. It should also be noted that P11 choose to fix the iron abundance of Fairall 9 at Fe/solar $= 2$, contrary to their approach for the other four AGN in their sample which have Fe/solar $= 1$.

The `compTT` soft excess used by P11 has a modest flux and optical depth, with $F_{0.6-10} = (3.6 \pm 0.2) \times 10^{-11} \, \mathrm{erg \, cm^{-2} \, s^{-1}}$ and $\tau = 0.5^{+1.6}_{0.2}$, but the upper limit on its temperature is quite high: $kT < 14.1 \, \mathrm{keV}$. This large temperature pushes the influence of this component almost into the Fe K band, possibly interfering with the measurement of the red wing of the broad Fe Kα line by the `kerrconv` inner disk

Fig. 4.20 Best-fitting model components for the 2007 *Suzaku* data fit by the L12 model including `compTT` emission for the soft excess. The total model is the *black solid line*, the power-law is the *black dashed line*, the inner disk reflector is in *blue*, the distant reflector in *magenta*, and the soft Comptonization component in *dashed red* (Figure is from Lohfink et al. (2012). Reproduced by permission of the AAS)

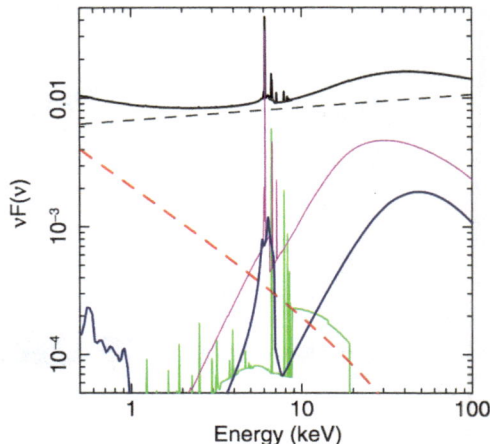

reflection model. Adopting these assumptions, the authors cannot constrain the spin of the SMBH in Fairall 9 using the dual reflector model, though they do achieve constraints using a more phenomenological approach by modeling the broad Fe Kα line alone with a `kerrdisk` component: $a = +0.67^{+0.10}_{-0.11}$.

A similar approach was adopted by L12 to examine modeling degeneracies in the soft excess (see Fig. 4.20). Comparing the L12 and P11 models fit to the *Suzaku* data, the addition of the `compTT` component (here with comparable optical depth, but higher temperature: $kT \sim 25\,\mathrm{keV}$) drives the iron abundance of the inner disk to a much higher value: Fe/solar $= 10^{+0}_{-2}$. The ionization of the inner disk is also much smaller ($\xi \leq 70\,\mathrm{erg\,cm\,s^{-1}}$), since the inner disk no longer has to account for the soft excess emission on its own. The inclination angle of the disk rises to $(48^{+6}_{-2})°$, while the spin of the black hole drops to $a = +0.52^{+0.19}_{-0.15}$.

In order to constrain the proper physical components of the spectrum, we must await the high S/N achievable by *NuSTAR*, which will be able to differentiate between the `compTT` (P11) and reflection-only (L12) models for the first time (see Fig. 4.21). Given the differences between the various modeling approaches used in L12 and P11, however, it is somewhat surprising that the spin constraints achieved in each case are consistent, within errors. This could be an indication that the presence of warm absorption is the greatest complicating factor in measuring spin, due to the curvature it induces in the spectrum interfering with the isolation of the red wing of the broad Fe Kα line. Alternatively (or perhaps in addition to this point), the nature of the spin parameter space could be playing a role in the similarity of the two measurements. As seen in Fig. 2.1, the shape of the function relating spin to the ISCO radius changes quite slowly and nearly linearly at moderate spin values below $a \leq 0.9$, but changes much more rapidly above $a \geq 0.9$. Therefore, differentiating between a spin of, e.g., $a = 0.4$ and $a = 0.7$ is much more challenging, statistically, then differentiating between spins of $a = 0.9$ and $a = 0.95$.

Fig. 4.21 A comparison of the two best-fitting models to the 2007 *Suzaku* data for Fairall 9 presented in L12: the reflection-only model (*blue*; lower line above 10 keV) and the reflection plus Comptonization model (*red*; upper line above 10 keV). Note the divergence of the two models above 10 keV, particularly. This divergence will be detectable with the high S/N spectra obtained from *NuSTAR* (Figure is from Lohfink et al. (2012). Reproduced by permission of the AAS)

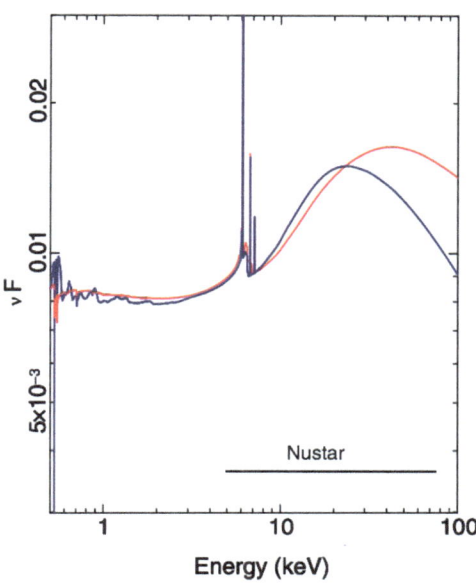

This point was discussed at some length in Walton et al. (2013; W13) as the authors analyzed the X-ray spectra of 25 "bare" Seyfert AGN with *Suzaku*, specifically to avoid the spectral complexities introduced by the presence of warm absorbing gas along the line of sight to the nucleus. In contrast to Patrick et al. (2012; P12), in which a Comptonized soft excess is assumed for the best fit to six bare Seyferts, W13 do not employ a separate model component for the soft excess, but rather allow it to be fit by the inner disk reflector. Of the five sources common to both the W13 and P12 samples, consistent spin measurements (within errors) were found for Mrk 335 and Ark 120, but not for Fairall 9, MCG–2-14-9 or NGC 7469. The sixth source from P12, SWIFT J2127.4+5654, was not considered in the W13 sample, but the spin derived in P12 is consistent with that measured by Miniutti et al. (2009a), who employed a reflection modeling scheme for the soft excess similar to W13. The model used to parametrize the soft excess has a noticeable impact on the derived spins in half of the small sample of P12 bare Seyferts, then, underscoring the importance of establishing the correct form of this spectral component. It is also interesting to note, however, that the three sources with consistent spin measurements in spite of modeling differences all have spins of $a \leq 0.85$, in keeping with the importance of the shape of the spin parameter space.

Chapter 5
Measuring the SMBH Spin Distribution

5.1 Sources of Systematic Error

In previous chapters we have noted the importance of both adequate data (i.e., high
S/N) and a physically self-consistent modeling approach to constraining SMBH
spins in AGN. We have also stressed the importance of one very critical assumption
that must be made in order to calculate black hole spin: namely, that the inner edge
of the accretion disk truncates at the ISCO. If the optically-thick disk is truncated
further out, then any spin derived using this assumption and the reflection modeling
technique will be a lower limit. Such truncated disks may reside in radio-loud AGN;
a disruption of the inner accretion flow (using dips in the X-ray light curve as a
proxy) is thought to coincide roughly with the ejection of new "knots" of plasma into
the radio jet (as observed in VLBI data). This behavior has been noted previous in,
e.g., 3C 120 (Marscher et al. 2002) and 3C 111 (Kataoka et al. 2007). By contrast, if
there is significant emission produced inside the ISCO, this will lead to a systematic
error on the black hole spin measurement obtained via the reflection method that
can be $\geq 20\%$ above the actual value of spin for non-spinning or retrograde black
holes, but is $\leq 2\%$ higher than the real spin for black holes with spins $a \geq +0.9$
(Reynolds and Fabian 2008).

The models currently used to represent both the accretion disk and the relativistic
smearing also have their inherent limitations and uncertainties. The `reflionx` and
`xillver` models both assume that the disk is thin and can be well-approximated by
a Novikov-Thorne formalism. This is still an active topic of debate: disk thickness
(e.g., Noble et al. 2011; Penna et al. 2010) and disk warping (e.g., Fragile and
Anninos 2005) at small radii can have a substantial impact on measurements of
black hole spin. The current models also assume that the disk has a constant density
and ionization structure throughout, which cannot be the case, physically. Portraying
them as such is a necessary simplification, computationally, and it is unclear whether
even the highest-quality data can differentiate between these simplified assumptions
and more complex models that have density and ionization varying as a function of
radius and/or depth in the disk.

L. Brenneman, *Measuring the Angular Momentum of Supermassive Black Holes*,
SpringerBriefs in Astronomy, DOI 10.1007/978-1-4614-7771-6_5,
© Laura Brenneman 2013

There is also some question about whether a limb-brightening vs. limb-darkening algorithm should be used to represent the directionality of the reflected emission from the disk when convolved with the smearing kernel (Svoboda et al., 2010). The nature of the disk emissivity profile itself is also an active topic of research; though the disk is thought to dissipate energy as a function of radius ($\epsilon \propto r^{-q}$), the emissivity index likely varies as a function of radius as well (Wilkins and Fabian 2011). The directionality of the coronal emission irradiating the disk also impacts the observed reflection spectrum. If the coronal photons reflected back onto the disk are produced from a compact, localized spot near the black hole spin axis and close to the disk surface, light bending effects will focus the coronal emission preferentially toward the center of the disk, resulting in an apparent enhancement of the disk emissivity at small radii (corresponding to $q \geq 3$). The degree to which the emission is centrally concentrated in this scenario depends on the height h of the coronal active region above the disk (e.g., Dauser et al. 2013; Miniutti and Fabian 2004). We currently lack the ability to characterize the physical properties of the corona in a given AGN, however, which limits our ability to understand disk irradiation and emissivity independently.

Finally, when multiple detectors are involved in collecting the data used to measure black hole spin, the cross-calibration uncertainty between detectors can also contribute to the systematic error on the spin constraint. Given the need for high-S/N spectra across a wide bandpass in X-rays, the use of multiple detectors is increasingly necessary in order to achieve a reliable spin measurement. For example, B11 and W13 discuss how uncertainty in the cross-calibration between the *Suzaku*/XIS and PIN instruments can affect constraints on spin in NGC 3783 and a sample of 25 bare Seyfert AGN, respectively. The *NuSTAR* mission, too, is currently working to improve the calibration between its two identical focal plane detectors, as well as its cross-calibration with the instruments on both *XMM-Newton* and *Suzaku*.

5.2 The Current Spin Sample

Taking all the caveats of Sect. 5.1 into account, one can begin to appreciate the challenge involved in obtaining precise, accurate spin constraints, and the limitations of our sample size to bright, nearby AGN that are relatively unobscured. For these reasons, there are currently only 22 AGN with robust, published constraints for their SMBH spins. Here, I have defined a "robust" constraint in a manner similar to that of Reynolds (2013), requiring that all other parameters of the accretion disk be left free to vary during the fit (i.e., the disk inclination angle, iron abundance and emissivity index, which must be itself constrained to $q \geq 2$ in order for the majority of the X-ray reflection to originate in the inner disk). The spins presented here meet these criteria and are single-valued. These AGN, and their properties, are listed in Table 5.1, and the histogram of spin values is shown in Fig. 5.1.

Table 5.1 Summary of black hole spin measurements derived from relativistic reflection fitting of SMBH X-ray spectra. All errors are quoted with 90% confidence for one interesting parameter (Data are taken with *Suzaku* except for 1H0707–495, which was observed with *XMM-Newton*; MCG–6-30-15, in which the data from *XMM-Newton* and *Suzaku* are consistent with each other; and NGC 1365, which was taken simultaneously with *XMM-Newton* and *NuSTAR*. Spin (a) is dimensionless, as defined previously. M is the mass of the black hole in solar masses, and $L_{\rm bol}/L_{\rm Edd}$ is the Eddington ratio of its luminous output. Host denotes the galaxy host type. All masses through 3C 120 are from Peterson et al. (2004) except MCG–6-30-15, 1 H0707–495 and SWIFT J2127.4+5654, which are taken from McHardy et al. (2005), Zoghbi et al. (2010) and Malizia et al. (2008), respectively. All bolometric luminosities of these same objects are from Woo and Urry (2002) except for the same three sources. The same references for MCG–6-30-15 and SWIFT J2127.4+5654 are used, but host types for 1H0707–495 and SWIFT J2127.4+5654 are unknown. Masses (bolometric luminosities) of the sources starting with 1 H0419–577 are from, respectively: Fabian et al. (2005) (same), Collier et al. (2001) (Romano et al. 2004) Gliozzi et al. (2010) (same), Turner et al. (2002) (same), Miniutti et al. (2009b) (Grupe et al. 2004), Zhou and Wang (2005) (same), Zhou and Wang (2005) (same), Kara et al. (2013) (Czerny et al. 2001), Bennert et al. (2011) (Woo and Urry 2002), Zhou and Wang (2005) (same), Risaliti et al. (2009b) (Vasudevan et al. 2010). References for each source are listed at the end of this work)

AGN	a	log M	$L_{\rm bol}/L_{\rm Edd}$	Host
MCG–6-30-15[a]	$\geq +0.98$	$6.65^{+0.17}_{-0.17}$	$0.40^{+0.13}_{-0.13}$	E/S0
Fairall 9[b]	$+0.52^{+0.19}_{-0.15}$	$8.41^{+0.11}_{-0.11}$	$0.05^{+0.01}_{-0.01}$	Sc
SWIFT J2127.4+5654[c]	$+0.6^{+0.2}_{-0.2}$	$7.18^{+0.07}_{-0.07}$	$0.18^{+0.03}_{-0.03}$	–
1 H0707–495[d]	$\geq +0.98$	$6.70^{+0.40}_{-0.40}$	$\sim 1.0_{-0.6}$	–
Mrk 79[e]	$+0.7^{+0.1}_{-0.1}$	$7.72^{+0.14}_{-0.14}$	$0.05^{+0.01}_{-0.01}$	SBb
Mrk 335[f]	$+0.70^{+0.12}_{-0.01}$	$7.15^{+0.13}_{-0.13}$	$0.25^{+0.07}_{-0.07}$	S0a
NGC 3783[g]	$\geq +0.98$	$7.47^{+0.08}_{-0.08}$	$0.06^{+0.01}_{-0.01}$	SB(r)ab
Ark 120[h]	$+0.94^{+0.1}_{-0.1}$	$8.18^{+0.05}_{-0.05}$	$0.04^{+0.01}_{-0.01}$	Sb/pec
3C 120[i]	≥ 0.95	$7.74^{+0.20}_{-0.22}$	$0.31^{+0.20}_{-0.19}$	S0
1 H0419–577[j]	$\geq +0.88$	$8.18^{+0.12}_{-0.12}$	$1.27^{+0.42}_{-0.42}$	–
Ark 564[j]	$+0.96^{+0.01}_{-0.06}$	≤ 6.90	≥ 0.11	SB
Mrk 110[j]	$\geq +0.99$	$7.40^{+0.09}_{-0.09}$	$0.16^{+0.04}_{-0.04}$	–
SWIFT J0501.9-3239[j]	$\geq +0.96$	–	–	SB0/a(s) pec
Ton S180[j]	$+0.91^{+0.02}_{-0.09}$	$7.30^{+0.60}_{-0.40}$	$2.15^{+3.21}_{-1.61}$	–
RBS 1124[j]	$\geq +0.98$	8.26	0.15	–
Mrk 359[j]	$+0.66^{+0.30}_{-0.54}$	6.04	0.25	pec
Mrk 841[j]	$\geq +0.52$	7.90	0.44	E
IRAS 13224-3809[j]	$\geq +0.995$	7.00	0.71	–
Mrk 1018[j]	$+0.58^{+0.36}_{-0.74}$	8.15	0.01	S0
IRAS 00521-7054[l]	$\geq +0.84$	–	–	–
NGC 4051[m]	$\geq +0.99$	6.28	0.03	SAB(rs)bc
NGC 1365[k]	$+0.97^{+0.01}_{-0.04}$	$6.60^{+1.40}_{-0.30}$	$0.06^{+0.06}_{-0.04}$	SB(s)b

[a] BR06, Miniutti et al. (2007)
[b] Schmoll et al. (2009), P12, L12, W13
[c] Miniutti et al. (2009a), P12
[d] Zoghbi et al. (2010), de le Calle Pérez et al. (2010), W13
[e] Gallo et al. (2005, 2011)
[f] P12, W13
[g] B11, P11
[h] P12, Nardini et al. (2011)
[i] Lohfink et al. (2013b)
[j] W13
[k] Risaliti et al. (2009b, 2013) and Brenneman et al. (2013)
[l] Tan et al. (2012)
[m] P12

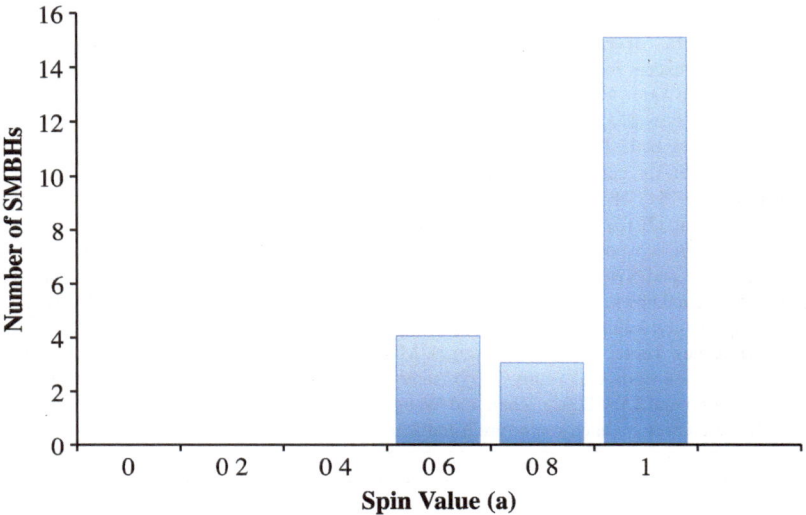

Fig. 5.1 Distribution of the 20 SMBH spins measured so far, based on data from Table 5.1. Note the weighting toward large prograde spin values

While it is difficult to draw any robust statistical inferences from a sample size of 22 objects, the trend toward higher spin values is obvious, even considering that systematic error (which can lower measured spins as described above in Sect. 5.1) is not taken into account in the uncertainties quoted here. There may be selection biases in play which may make it more likely that we measure higher spin values: AGN whose disks extend down closer to the event horizon (i.e., those with large, prograde SMBH spins) accrete more efficiently than those whose disks truncate farther from the SMBH, provided that the disks in question conform to standard thin-disk profiles (e.g., Novikov-Thorne). As such, an accreting, rapidly spinning black hole will be more luminous than an accreting, slowing spinning black hole and hence will be over-represented in flux-limited samples (B11). The nature of the spin parameter space may also be playing a role here, as discussed in W13: because of the rapid change in the shape of the spin function vs. the ISCO radius at large prograde spin values, it is easier to constrain spins with greater precision and accuracy when they have spin values closer to $a = +1$.

Nonetheless, the pattern that is most readily apparent in Table 5.1 is that 15/22 AGN have relatively high, prograde SMBH spins ($a \geq 0.8$), and no retrograde spins have conclusively been measured (although the 90% confidence lower bound on the spin of the SMBH in of Mrk 1018 allows for retrograde spin). Cowperthwaite and Reynolds (2012) previously published a spin constraint of $a \leq -0.1$ for 3C 120, but by taking multi-epoch, multi-wavelength data and the latest *Suzaku* calibrations into account, Lohfink et al. (2013b) have revised this measurement to $a \geq +0.95$.

3C 120 is the one radio-loud galaxy with a measured spin in the current sample, and is thus of great interest in terms of probing the connection between black hole

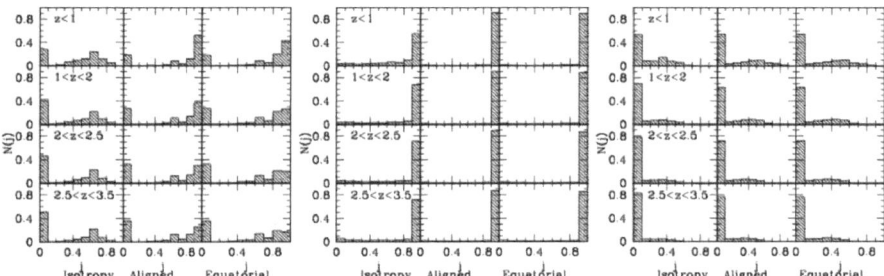

Fig. 5.2 Spin distribution as a function of redshift for the simulated SMBHs of Berti and Volonteri (2008). The *left plot* shows spin evolution driven by black hole mergers only, the *middle plot* shows mergers plus prolonged, prograde accretion, and the *right plot* shows mergers plus chaotic, random episodes of accretion. *Left*-to-*right* columns in each plot show isotropic, aligned and equatorially-oriented mergers, respectively (Figure is in Berti and Volonteri (2008). Reproduced by permission of the AAS)

spin and jet production. Garofalo (2009) postulated that jet power is maximized for rapidly-rotating retrograde black holes, though this idea is not without controversy (e.g., Tchekhovskoy and McKinney 2012). More work needs to be done to constrain black hole spin and jet power independently from observations in order to prove or disprove this conjecture, and to place the rapid prograde spin measured for 3C 120 in context with SMBH spins and jet luminosities for other radio-loud AGN. It is worth noting, however, that even the modest distribution in spin values seen in Table 5.1 implies that black hole spin cannot be the primary driver in determining whether an AGN possesses a relativistic jet.

Narayan and McClintock (2012) and Steiner et al. (2012) demonstrate two examples of the beginnings of such research in microquasars. These authors report a correlation between jet power and spin ($P_{\text{jet}} \propto a^2$) in a sample of five stellar-mass black holes, as expected based on the theoretical work of Blandford and Znajek (1977). There is some disagreement about this finding, however (e.g., see Fender et al. 2010), largely centered on how the jet power is measured. Daly (2011), meanwhile, has made a first effort at measuring SMBH spins in 55 radio-loud AGN, finding a distribution with an average close to $a = +0.5$, but with large uncertainties on the individual spin values. Precise measurements of AGN jet magnetic fields are necessary in order to definitively constrain the SMBH spins in these sources, however, and a $P_{\text{jet}} \propto a^2$ relation is assumed *a priori* in the work.

If the trend toward large prograde spins continues to hold as our sample size increases, we might ultimately infer that the growth of bright, nearby AGN in recent epochs has been driven primarily by prolonged, prograde accretion of gas. If the overall distribution of SMBH spins in the local universe begins to drift toward intermediate values, it is likely that the role of mergers has been more significant than that of ordered gas accretion. Similarly, if the distribution tends toward low values of spin, we can infer that episodes of randomly-oriented accretion have been the dominant means of SMBH and galaxy growth (Berti and Volonteri 2008; see

Fig. 5.2). Reynolds (2013) further note that both the most and least massive SMBHs in Table 5.1 seem to have more moderate spin values than their rapidly-spinning counterparts in the middle of the mass range. If this trend continues as the sample size of measured SMBH spins grows, it would provide direct evidence for the increased role of chaotic accretion and/or major mergers at these two extreme ends of the SMBH mass spectrum.

Chapter 6
Conclusions and Future Directions

Measuring black hole spin is painstaking work, even with the best data from current observatory-class missions such as *XMM-Newton*, *Suzaku* and *Chandra*. Long observations (∼hundreds of kiloseconds) of bright AGN are needed, and multi-epoch, multi-instrument data should be analyzed jointly whenever possible in order to assess the physical nature and variability of all of the components in a given X-ray spectrum, knowing that the SMBH spin value will not change over human timescales. High S/N across A broad energy range is also desirable in order to constrain the properties of the continuum and complex absorption, particularly, and to distinguish these components from any signatures of inner disk reflection. Only by isolating the broad Fe Kα line and its associated Compton hump can we measure black hole spin with the accuracy and precision necessary to begin constructing a spin distribution for local AGN. We can then begin to draw inferences regarding the dominant growth mechanism of these SMBHs over cosmic time, and to understand the role of spin in jet production and AGN feedback.

Our current sample of 22 AGN with measured, published SMBH spins must be extended in order to accomplish these goals. The *Suzaku* Spin Survey has recently been completed, and is providing rich legacy datasets that will benefit this science for years to come. Additionally, many datasets from the *XMM-Newton* and *Suzaku* archives have recently been analyzed with an eye toward measuring spin (e.g., P11, P12, W13). *NuSTAR* will also play a vital role in this science, providing an invaluable high-energy (∼3–80 keV) complement to *XMM-Newton* and *Suzaku* spectra, particularly, when used simultaneously with either observatory. This high-energy capability will improve the accuracy of black hole spin measurements, and will also enable improvements in precision in these measurements by up to a factor of 10 in some sources (e.g., Fig. 4.7).

Astro-H, scheduled for launch in 2015, will bring the science of micro-calorimetry to X-ray astronomy with a spectral resolution of $\Delta E \sim 7\,\mathrm{eV}$ over the 0.3–12 keV range. Though the observatory will also fly a high-energy detector capable of producing spectra up to 600 keV, the calorimeter will be the unique

L. Brenneman, *Measuring the Angular Momentum of Supermassive Black Holes*,   45
SpringerBriefs in Astronomy, DOI 10.1007/978-1-4614-7771-6_6,
© Laura Brenneman 2013

strength of this mission, enabling the broad and narrow Fe K emission and absorption features to be definitively disentangled and the telltale signatures of complex intrinsic absorption to be identified and modeled correctly.

In order to achieve the order of magnitude increase in sample size necessary to begin assessing the spin distribution of SMBHs in the local universe from a statistical perspective, future large-area ($\geq 1\,\mathrm{m}^2$) X-ray observatories are needed. Proposed concepts such as *IXO/AXSIO* (White et al. 2010), *ATHENA* (now ATHENA +; Barcons et al. 2012) and the *Extreme Physics Explorer (EPE)* (Garcia et al. 2011) would all offer the necessary collecting area and superior spectral resolution, allowing us to extend our sample of measured SMBH spins to several hundred AGN using the reflection modeling method.

Additionally, such large-area observatories will also enable the orbits of distinct "blobs" or "hot spots" of material within the accretion disk to be measured via the periodicity of their emission, allowing velocity to be charted as a function of radius within the disk for tens of AGN. Such measurements would provide an independent check on the spin value obtained from spectral fitting of the inner disk reflection signatures averaged over many orbits, and would also yield important constraints on black hole masses as well. The Large Observatory For Timing (*LOFT*; Feroci et al. 2012) is a proposed concept that, if funded, would provide the necessary effective area ($\geq 10\,\mathrm{m}^2$) to achieve this goal, coupled with moderate spectral resolution ($\leq 260\,\mathrm{eV}$) across a reasonably broad bandpass (2–30 keV). As discussed in Reynolds (2013), *LOFT* would also revolutionize the science of relativistic reverbertion mapping, in which the lag time is measured between variations in the continuum emission from the corona and variations in the response of the observed X-ray reflection from the inner accretion disk. The full, energy-dependent transfer function relating the changes in these two X-ray spectral signatures encodes the spin of the black hole, among other physical information about the inner accretion flow (Zoghbi and Fabian 2011, de Marco et al. 2011, Fabian et al. 2012, Kara et al. 2013). Having an instrument such as *LOFT* at our disposal would thus provide two more methods to use in determining black hole spins.

The science of measuring the angular momenta of black holes is in its infancy. Though the past decade has seen great strides in our ability to constrain spin through long X-ray observations coupled with detailed spectral modeling, much work remains to be done in terms of improving the precision and accuracy of these measurements, as well as the sample size. The next decade will see an improvement in the quality of black hole spin science, but a significant advance in the quantity of this work in the decades beyond will depend critically on the amount of funding available to facilitate the international collaborations necessary to build large-area X-ray spectroscopy missions, or on advances in technology development that will allow such a large-area X-ray spectroscopic mission to be flown for a fraction of the current cost.

Chapter 7
Epilogue: *NuSTAR* Validates Inner Disk Reflection in NGC 1365

The *NuSTAR* X-ray observatory has recently undertaken an ambitious campaign to observe several AGN simultaneously with either *XMM-Newton* or *Suzaku*. These deep observations will yield the highest S/N spectra from 0.2 to 80 keV ever obtained, enabling the continuum, absorption and reflection components of these AGN to be unambiguously disentangled. As discussed in Sect. 4.1, deconvolving these spectral features will allow black hole spin to be measured with greater precision and accuracy than has ever been achieved in previous work.

The Seyfert 1.8 AGN NGC 1365 is the only AGN known to display, in addition to the near-ubiquitous continuum and reflection from distant material, (1) extended X-ray emission from a circumnuclear starburst (Wang et al. 2009), (2) relativistic inner disk reflection (Risaliti et al. 2009a, Walton et al. 2010, Brenneman et al. 2013), (3) a warm absorber (Risaliti et al. 2005b, Brenneman et al. 2013), and (4) a time-variable cold absorber that eclipses the inner disk/corona (Risaliti et al. 2005a, Maiolino et al. 2010, Brenneman et al. 2013). It has been the subject of over a dozen X-ray observations with *XMM-Newton, Chandra* and *Suzaku* during the past decade. Recently, Brenneman et al. (2013) jointly analyzed *Suzaku* spectra from three different observations over a 2-year period in order to maximize S/N in an effort to separate the various spectral components. A preliminary spin constraint of $a = 0.96 \pm 0.01$ was obtained using a `relconv(reflionx)` model for the inner disk reflection. The limited S/N of the *Suzaku* data above 10 keV made it difficult to uniquely establish relativistic reflection as the best-fitting model, however (e.g., vs. multiple complex absorbers), calling into question the ability of the data to truly constrain spin.

NGC 1365 was one of the first AGN observed by *NuSTAR* as part of its science operations phase, and has now been the subject of four separate observations taken simultaneously with *XMM-Newton*. These observations were taken in July and December 2012, and in January and February 2013, and total nearly 500 ks of simultaneous data from the two telescopes. The data from all four observations are currently being analyzed, and the first results from spectral fitting of the July 2012 observation have now been published (Risaliti et al. 2013).

L. Brenneman, *Measuring the Angular Momentum of Supermassive Black Holes*,
SpringerBriefs in Astronomy, DOI 10.1007/978-1-4614-7771-6_7,
© Laura Brenneman 2013

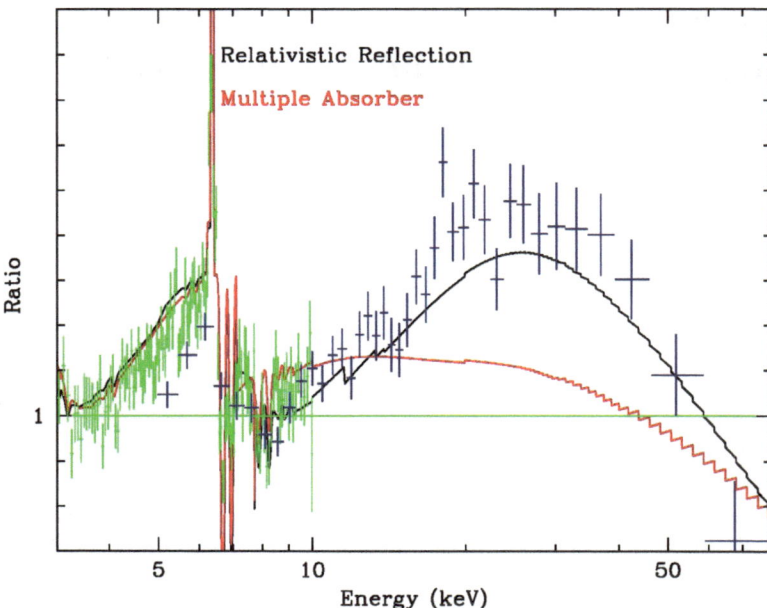

Fig. 7.1 The relativistic inner disk reflection model (*black line*) vs. the absorption-only model (*red line*) plotted against the *XMM-Newton* (*green*) and *NuSTAR* (*blue*) spectral data. Both models are fit only below 10 keV and extrapolated above this energy. Note the strong resemblance of the data to the reflection model and divergence from the absorption model at high energies. Both models fit the data equally well below 10 keV (Credit: G. Risaliti, private communication)

Just as simulations for MCG–6-30-15 predicted that the addition of *NuSTAR* data to that from *XMM-Newton* would enable the reflection and absorption-only models to be conclusively disentangled (Figs. 4.6 and 4.7), the early *NuSTAR+XMM-Newton* observations of NGC 1365 have conclusively demonstrated this capability. Figure 7.1 shows the reflection model (black line) and the absorption-only model (red line) fit to the July 2012 *XMM-Newton* data (green points) below 10 keV. The models are then extrapolated up to 79 keV and the *NuSTAR* data (blue points) are added, *without refitting*. Though the two models fit the data equally well below 10 keV, note the clear divergence of the two models above this energy, the striking agreement between the *NuSTAR* data and the reflection model, and the clear disagreement between the *NuSTAR* data and the absorption-only model. The data overwhelmingly support the presence of inner disk reflection signatures in the data, in addition to both cold and warm absorption. Applying a `relconv`(`reflionx`) model to the spectra, a spin constraint of $a = 0.97^{+0.01}_{-0.04}$ is obtained, as quoted in Table 5.1. The high S/N and broad-band spectral coverage of these data make this the most statistically accurate, precise spin constraint achieved to date. The robustness of this spin measurement is best appreciated through an examination of the change in statistical goodness-of-fit with spin value, as shown in Fig. 7.2.

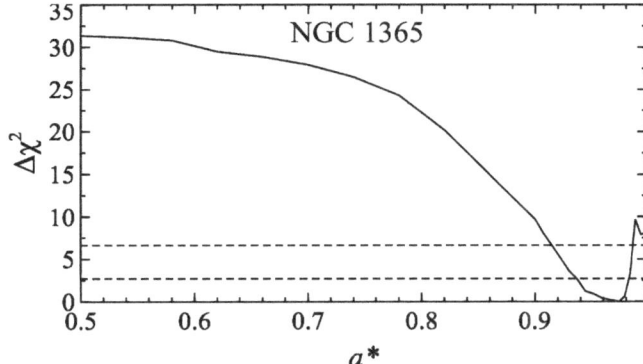

Fig. 7.2 Change in the global goodness-of-fit vs. spin for the reflection model, applied to data from the July 2012 NGC 1365 *NuSTAR+XMM-Newton* observation. The *dashed lines* represent the 99 % (*upper*) and 90 % (*lower*) confidence intervals (Figure is from Risaliti et al. (2013). Reprinted by permission from Macmillan Publishers, Ltd.)

Acknowledgements LB thanks Chris Reynolds, Martin Elvis, Mike Nowak, Jon Miller, Andy Fabian, Rubens Reis, Guido Risaliti, Dom Walton, Anne Lohfink, Emanuele Nardini, Giorgio Matt, Daniel Stern and Fiona Harrison for useful discussions and collaborations that have contributed to many of the spin measurements presented in this work.

References

Agol, E., Krolik, J.H.: Astrophys. J. **528**, 161 (2000)
Antonucci, R.: Annu. Rev. Astron. Astrophys. **31**, 473 (1993)
Arnaud, K., et al.: Mon. Not. R. Astron. Soc. **217**, 105 (1985)
Barcons, X., et al.: (2012). arXiv:1207.2745
Bardeen, J., Press, W., Teukolsky, S.: Astrophys. J. **178**, 347 (1972)
Beckwith, K., Done, C.: Mon. Not. R. Astron. Soc. **359**, 1217 (2005)
Bennert, V., et al.: Astrophys. J. **726**, 59 (2011)
Berti, E., Volonteri, M.: Astrophys. J. **684**, 822 (2008)
Bianchi, S., et al.: Astron. Astrophys. **495**, 421 (2009)
Blandford, R.D., McKee, C.: Astrophys. J. **255**, 419 (1982)
Blandford, R.D., Znajek, R.L.: Mon. Not. R. Astron. Soc. **179**, 133 (1977)
Brenneman, L., Reynolds, C.: Astrophys. J. **652**, 1028 (2006)
Brenneman, L., Reynolds, C.: Astrophys. J. **702**, 1367 (2009)
Brenneman, L., et al.: Astrophys. J. **736**, 103 (2011)
Brenneman, L., et al.: Astrophys. J. **744**, 13 (2012)
Brenneman, L., et al.: Mon. Not. R. Astron. Soc. **429**, 2662 (2013)
Broderick, A., et al.: Astrophys. J. **735**, 57 (2011)
Chayer, P., Fontaine, G., Wesemael, F.: Astrophys. J. Suppl. **99**, 189 (1995)
Chiang, C.Y., Fabian, A.: Mon. Not. R. Astron. Soc. **414**, 2345 (2011)
Collier, S., et al.: Astrophys. J. **561**, 146 (2001)
Coppi, P.: ASP Conf. Ser. **161**, 375 (1999)
Cowperthwaite, P., Reynolds, C.: Astrophys. J. **752L**, 21 (2012)
Crummy, J., et al.: Mon. Not. R. Astron. Soc. **365**, 1067 (2006)
Czerny, B., et al.: Mon. Not. R. Astron. Soc. **325**, 865 (2001)
Daly, R.: Mon. Not. R. Astron. Soc. **414**, 1253 (2011)
Dauser, T., et al.: Mon. Not. R. Astron. Soc. **409**, 1534 (2010)
Dauser, T., et al.: Mon. Not. R. Astron. Soc. (2013). arXiv:1301.4922
Davis, S., et al.: Astrophys. J. **647**, 525 (2006)
de La Calle Pérez, I., et al.: Astron. Astrophys. **524**, 50 (2010)
de Marco, B., et al.: Mon. Not. R. Astron. Soc. **417**, L98 (2011)
Di Matteo, T.: Am. Inst. Phys. Conf. **599**, 83 (2001)
Doeleman, S., et al.: Nature **455**, 78 (2008)
Done, C., et al.: Mon. Not. R. Astron. Soc. **420**, 1848 (2012)
Dovčiak, M., Karas, V., Yaqoob, T.: Astrophys. J. Suppl. **153**, 205 (2004)
Elvis, M.: (2012). arXiv:1201.3520
Emmanoulopoulos, D., et al.: Mon. Not. R. Astron. Soc. **415**, 1895 (2011)

L. Brenneman, *Measuring the Angular Momentum of Supermassive Black Holes*,
SpringerBriefs in Astronomy, DOI 10.1007/978-1-4614-7771-6,
© Laura Brenneman 2013

Fabian, A.C.: Annu. Rev. Astron. Astrophys. **50**, 455 (2012)
Fabian, A.C., et al.: Mon. Not. R. Astron. Soc. **238**, 729 (1989)
Fabian, A.C., et al.: Mon. Not. R. Astron. Soc. **335L**, 1 (2002)
Fabian, A.C., et al.: Mon. Not. R. Astron. Soc. **361**, 795 (2005)
Fabian, A.C., et al.: Mon. Not. R. Astron. Soc. **419**, 116 (2012)
Fender, R., Gallo, E., Russell, D.: Mon. Not. R. Astron. Soc. **406**, 1425 (2010)
Ferland, G.J., et al.: Rev. Mex. Astron. Astrophys. **49**, 1 (2013)
Feroci, M., et al.: Exp. Astron. **34**, 415 (2012)
Ferrarese, L., Merritt, D.: Astrophys. J. **539L**, 9 (2000)
Fragile, C.P., Anninos, P.: Astrophys. J. **623**, 347 (2005)
Frank, J., King, A., Raine, D.J.: Accretion Power in Astrophysics: Third Edition. University Press, Cambridge, U.K. (2002)
Gallo, L.C., et al.: Mon. Not. R. Astron. Soc. **363**, 64 (2005)
Gallo, L.C., et al.: Mon. Not. R. Astron. Soc. **411**, 607 (2011)
Garcia, J., Kallman, T.: Astrophys. J. **718**, 695 (2010)
Garcia, M., et al.: Soc. Photo-Opt. Instrum. Eng. **8147E**, 55 (2011)
Garofalo, D.: Astrophys. J. **699**, 400 (2009)
Genzel, R., et al.: Mon. Not. R. Astron. Soc. **317**, 348 (2000)
George, I., Fabian, A.: Mon. Not. R. Astron. Soc. **249**, 352 (1991)
Ghez, A., et al.: Nature **407**, 349 (2000)
Gierlinski, M., et al.: Nature **455**, 369 (2008)
Gliozzi, M., et al.: Astrophys. J. **717**, 1243 (2010)
Grupe, D., et al.: Astron. J. **127**, 156 (2004)
Guainazzi, M., et al.: Astron. Nachr. **327**, 1032 (2006)
Gültekin, K., et al.: Astrophys. J. **698**, 198 (2009)
Halpern, J.: Astrophys. J. **281**, 90 (1984)
Harrison, F.A., et al.: Astrophys. J. **770**, 103 (2013)
Hawking, S.W.: Nature **248**, 30 (1974)
Kallman, T., Bautista, M.: Astrophys. J. Suppl. **133**, 221 (2001)
Kara, E., et al.: Mon. Not. R. Astron. Soc. **430**, 1408 (2013)
Kataoka, J., et al.: Publ. Astron. Soc. Jpn. **59**, 279 (2007)
Kerr, R.P.: Phys. Rev. Lett. **11**, 237 (1963)
Krongold, Y., et al.: Astrophys. J., **597**, 832 (2003)
Krongold, Y., et al.: Am. Inst. Phys. Conf. **774**, 325 (2005)
Laor, A.: Astrophys. J. **376**, 90 (1991)
Lee, J.C.: Space Sci. Rev. **157**, 93 (2010)
Lohfink, A., et al.: Astrophys. J. **758**, 67 (2012)
Lohfink, A., et al.: (2013a). arXiv:1301.4997
Lohfink, A., et al.: (2013b). arXiv:1305.4937
Magdziarz, P., Zdziarski, A.A.: Mon. Not. R. Astron. Soc. **273**, 837 (1995)
Maiolino, R., et al.: Astron. Astrophys. **517A**, 47 (2010)
Malizia, A., et al.: Mon. Not. R. Astron. Soc. **389**, 1360 (2008)
Markoff, S., Nowak, M., Wilms, J.: Astrophys. J. **635**, 1203 (2005)
Marscher, A.P., et al.: Nature **417**, 625 (2002)
Matt, G., et al.: Astron. Astrophys. **257**, 63 (1992)
McHardy, I., Papadakis, I., Uttley, P.: Nucl. Phys. B. Proc. Supp. **69**, 509 (1999)
McHardy, I., et al.: Mon. Not. R. Astron. Soc. **359**, 1469 (2005)
Miller, J.: Annu. Rev. Astron. Astrophys. **45**, 441 (2007)
Miller, M., Colbert, E.: Int. J. Mod. Phys. **13**, 1 (2004)
Miller, L., Turner, T., Reeves, J.: Astron. Astrophys. **483**, 437 (2008)
Miller, L., Turner, T., Reeves, J.: Mon. Not. R. Astron. Soc. **399L**, 69 (2009)
Miniutti, G., Fabian, A.: Mon. Not. R. Astron. Soc. **349**, 1435 (2004)
Miniutti, G., et al.: Publ. Astron. Soc. Jpn. **59S**, 315 (2007)
Miniutti, G., et al.: Mon. Not. R. Astron. Soc. **398**, 255 (2009a)

Miniutti, G., et al.: Mon. Not. R. Astron. Soc. **401**, 1315 (2009b)
Murphy, K., Yaqoob, T.: Mon. Not. R. Astron. Soc. **397**, 1549 (2009)
Nandra, K., et al.: Mon. Not. R. Astron. Soc. **382**, 194 (2007)
Narayan, R., McClintock, J.: Mon. Not. R. Astron. Soc. **419L**, 69 (2012)
Nardini, E., et al.: Mon. Not. R. Astron. Soc. **410**, 1251 (2011)
Noble, S.C., et al.: Astrophys. J. **743**, 115 (2011)
Patrick, A., et al.: Mon. Not. R. Astron. Soc. **416**, 2725 (2011)
Patrick, A., et al.: Mon. Not. R. Astron. Soc. **411**, 2353 (2012)
Penna, R.F., et al.: Mon. Not. R. Astron. Soc. **408**, 752 (2010)
Peterson, B., et al.: Astrophys. J. **613**, 682 (2004)
Pétri, J.: Astrophys. Space Sci. **318**, 181 (2008)
Poutanen, J., Svensson, R.: Astrophys. J. **470**, 249 (1996)
Reis, R., et al.: Astrophys. J. **745**, 93 (2012)
Remillard, R., McClintock, J.: Annu. Rev. Astron. Astrophys. **44**, 49 (2006)
Reynolds, C.S.: Mon. Not. R. Astron. Soc. **286**, 513 (1997)
Reynolds, C.S.: (2013). arXiv:1302.3260
Reynolds, C.S., Fabian, A.: Astrophys. J. **679**, 1181 (2008)
Reynolds, C.S., Nowak, M.: Phys. Rep. **377**, 389 (2003)
Reynolds, C.S., et al.: Astrophys. J. **755**, 88 (2012)
Risaliti, G., et al.: Astrophys. J. **623L**, 93 (2005a)
Risaliti, G., et al.: Astrophys. J. **630L**, 129 (2005b)
Risaliti, G., et al.: Mon. Not. R. Astron. Soc. **393L**, 1 (2009a)
Risaliti, G., et al.: Astrophys. J. **696**, 160 (2009b)
Risaliti, G., et al.: Nature **494**, 449 (2013)
Romano, P., et al.: Astrophys. J. **602**, 635 (2004)
Ross, R., Fabian, A.: Mon. Not. R. Astron. Soc. **358**, 211 (2005)
Schmoll, S., et al.: Astrophys. J. **703**, 2171 (2009)
Schnittman, J., Krolik, J.: Astrophys. J. **701**, 1175 (2009)
Seaton, M.: Astrophys. Space Sci. **237**, 107 (1996)
Shakura, N., Sunyaev, R: Astron. Astrophys. **24**, 337 (1973)
Silvestro, G.: Astron. Astrophys. **36**, 41 (1974)
Steiner, J., et al.: Astrophys. J. **745**, 136 (2012)
Strohmayer, T.: Astrophys. J. **552L**, 49 (2001)
Svoboda, J., et al.: Am. Inst. Phys. Conf. **1248**, 515 (2010)
Takahashi, T., et al.: Soc. Photo-Opt. Instrum. Eng. **7732E**, 27 (2010)
Tan, Y., et al.: Astrophys. J. **747**, L11 (2012)
Tanaka, Y., et al.: Nature **375**, 659 (1995)
Tchekhovskoy, A., McKinney, J.: Mon. Not. R. Astron. Soc. **423L**, 55 (2012)
Thorne, K.: Astrophys. J. **191**, 507 (1974)
Titarchuk, L.: Astrophys. J. **434**, 313 (1994)
Tombesi, F., et al.: Astron. Astrophys. **521A**, 57 (2010)
Tomsick, J., et al.: (2009). arXiv:0902.4238
Turner, T.J., et al.: Astrophys. J. **568**, 120 (2002)
Urry, C., Padovani, P.: Publ. Astron. Soc. Pac. **107**, 803 (1995)
Uttley, P., McHardy, I., Vaughan, S.: Mon. Not. R. Astron. Soc. **359**, 345 (2005)
Vasudevan, R.V., et al.: Mon. Not. R. Astron. Soc. **402**, 1081 (2010)
Volonteri, M., et al.: Astrophys. J. **620**, 69 (2005)
Walton, D.J., Reis, R.C., Fabian, A.C.: Mon. Not. R. Astron. Soc. **408**, 601 (2010)
Walton, D.J., et al.: Mon. Not. R. Astron. Soc. **428**, 2901 (2013)
Wang, J., et al.: Astrophys. J. **694**, 718 (2009)
Wassermann, D., et al.: Astron. Astrophys. **524A**, 9 (2010)
Watson, W., Wallin, B.: Astrophys. J. **432L**, 35 (1994)
White, N., et al.: Am. Inst. Phys. Conf. **1248**, 561 (2010)
Wilkins, D., Fabian, A.: Mon. Not. R. Astron. Soc. **414**, 1269 (2011)

Wilms, J., Allen, A., McCray, R.: Astrophys. J. **542**, 914 (2000)
Woo, J.H., Urry, C.M.: Astrophys. J. **579**, 530 (2002)
Yaqoob, T., Padmanabhan, U.: Astrophys. J. **604**, 63 (2004)
Zhou, X.-L., Wang, J.-M.: Astrophys. J. **618**, L83 (2005)
Zoghbi, A., Fabian, A.C.: Mon. Not. R. Astron. Soc. **418**, 2642 (2011)
Zoghbi, A., et al.: Mon. Not. R. Astron. Soc. **401**, 2419 (2010)